Fluid Flow in Discontinuous Rocks

Fluid Flow in Discontinuous Rocks

Cheng-Haw Lee

and

Ian Farmer

CHAPMAN & HALL
London · Glasgow · New York · Tokyo · Melbourne · Madras

Published by Chapman & Hall, 2-6 Boundary Row, London SE1 8HN

Chapman & Hall, 2-6 Boundary Row, London SE1 8HN, UK

Blackie Academic & Professional, Wester Cleddens Road, Bishopbriggs, Glasgow G64 2NZ, UK

Chapman & Hall Inc., 29 West 35th Street, New York NY10001, USA

Chapman & Hall Japan, Thomson Publishing Japan, Hirakawacho Nemoto Building, 6F, 1-7-11 Hirakawa-cho, Chiyoda-ku, Tokyo 102, Japan

Chapman & Hall Australia, Thomas Nelson Australia, 102 Dodds Streets, South Melbourne, Victoria 3205, Australia

Chapman & Hall India, R. Seshadri, 32 Second Main Road, CIT East, Madras 600 035, India

First edition 1993

Typeset in 10/12 Palatino by Thomson Press (India) Ltd, New Delhi

Printed in Great Britain by T. J. Press, Padstow, Cornwall

ISBN 0 412 41510 0

A catalogue record for this book is available from the British Library

Library of Congress Cataloging-in-Publication data available

∞ Printed on permanent acid-free text paper, manufactured in accordance with the proposed ANSI/NISO Z 39.48-199X and ANSI Z 39.48-1984

Contents

Preface

Rock masses comprise systems of discontinuities developed by tectonic action or diagenesis, and occasionally imposed by excavation, which bound blocks of intact rock. Since intact rock – with exceptions – has limited ability to transmit fluids most fluid flow in rock masses is through the apertures of the discontinuity systems. Thus, fluid flow is determined by the distribution of discontinuities and the way in which their apertures respond to natural and imposed stress fields in the rock mass.

The traditional approach to analysis assumes an equivalent continuum representing the discontinuous rock mass with permeabilities determined by borehole tests sampling a limited volume of the mass. This has limitations, particularly in rocks with variable or widely–spaced discontinuities, where it is difficult to sample a representative volume. More modern approaches incorporate discontinuity parameters attempting to incorporate the flow characteristics of individual discontinuities into a model of a connected discontinuity system. These are limited by difficulties in defining or measuring discontinuity systems and apertures.

The authors aim to examine both of these approaches but the principal aim is to develop simple usable fracture flow models based on an understanding of discontinuity structures and the mechanics of rock and discontinuity deformation, and the incorporation of these into simple models developed from percolation theory.

The book aims to be introductory, but the authors also hope that it can be used by engineers – particularly those designing underground structures where inflow is a problem. The authors are engineers and their principal interest is in the behaviour of rock masses. Their approach is conditioned by the basic assumption that the most important factors affecting a correct interpretation of fluid flow in rock masses are an accurate description of the rock mass structure and a knowledge of the deformation characteristics of rock.

The authors acknowledge a very considerable debt to a former and much missed colleague, Philip Garritty, and are grateful for the help – passive and active – of former colleagues at the University of Arizona.

Rock and rock mass permeability 1

It is useful as a preliminary step to look at some of the simpler approaches to groundwater flow and to record some of the data available on fluid flow through rocks. A rock mass can most simply be described as a series of blocks of intact rock separated by discontinuities. Fluid flow through the rock mass is determined both by the properties of the intact rock and of the discontinuities. In this it differs from conventional Darcian flow which assumes a homogeneous porous continuum with flow through interconnected pore spaces. In practice, and especially in crystalline rocks, the intact rock usually has such low permeability that discontinuities are dominant and fluid flow occurs mainly through discontinuities. Where this occurs, the hydraulic behaviour of the rock mass is determined by the geometry of the fracture or discontinuity system.

In order to analyse the hydraulic behaviour of rock masses, fracture flow models are used to simulate discontinuous rock for the purpose of predicting flow. These flow models can be divided into two groups: continuum models and discrete network models.

In **continuum** models, the discontinuous medium is assumed to have a sufficient number of randomly orientated and interconnected fractures to make it possible to define its average properties statistically and meaningfully. Thus heterogeneous fractured media are treated as homogeneous porous media (Long *et al.*, 1982; Shapiro and Andersson, 1983; Long, 1985). Parameters such as permeability and porosity can then be estimated, ideally from field tests, although because of cost and complexity, data are not always readily available.

In **discrete network** models, each fracture is modelled individually or a single equivalent fracture can replace several related fractures. This type of approach allows the influence of discontinuity parameters on flow to be examined. The choice of model depends on the amount of geologic information available and the volume and heterogeneity of the rock mass.

1.1 PERMEABILITY

Permeability is a measure of the ability of a medium to transmit fluid through its connected pore space. Like electrical or thermal conductivity, permeability is a second rank tensor relating a flux to a gradient:

$$q_i = -\frac{k_{ij}}{\mu}\frac{\delta P}{\delta x_j} \tag{1.1}$$

where q_i is the specific discharge, $\delta P/\delta x_j$ is the pressure gradient causing flow, μ is the fluid viscosity and k_{ij} are permeability constants. This relation has an essentially empirical derivation and is known as Darcy's Law. It is mathematically equivalent to Ohm's Law and Fick's First Law, respectively describing electrical and thermal conductivity.

Although permeability is frequently highly directional, principal values of k_{ij} are only rarely available and it generally becomes necessary to treat permeability as a scalar quantity. Equation 1.1 then reduces to:

$$q = -\frac{k}{\mu}\frac{\delta P}{\delta x}. \tag{1.2}$$

The **permeability coefficient**, or intrinsic permeability, k, is independent of the properties of the permeating fluid; depending solely on the structural characteristics of the medium. It has dimension L^2 and a unit of a darcy ($\cong 10^{-12}\,\mathrm{m}^2$).

In circumstances where the percolating fluid is water it is often convenient to define a coefficient of hydraulic conductivity which reflects both fluid and medium properties. An alternative form of equation 1.2 is then:

$$q = -\frac{K}{\gamma_f}\frac{\delta P}{\delta x} \tag{1.3}$$

where γ_f is the fluid specific weight; or more generally:

$$Q = AK\frac{\delta h}{\delta l} \tag{1.4}$$

where Q is the discharge flow rate (L^3T^{-1}), A is the gross discharge area (L^2) and $\delta h/\delta l$ is the dimensionless **hydraulic gradient** usually represented by i.

This is the commonest form of Darcy's Law and expresses permeability in terms of the **coefficient of hydraulic conductivity**, K, which has dimensions LT^{-1} and is related to the intrinsic permeability coefficient by the expression:

$$K = \frac{k\gamma_f}{\mu}, \tag{1.5}$$

thus for water, 1 darcy is approximately equal to $10^{-5}\,\mathrm{ms}^{-1}$.

The specific discharge or Darcian flow velocity given by equation 1.4 is calculated assuming, in effect, that flow occurs through the gross cross-sectional area of the medium. In reality flow takes place through the average interconnected pore area and the average flow velocity through the pores of the medium, termed the seepage velocity, is therefore larger than the Darcian velocity. The seepage velocity is approximated by the expression:

$$v = \frac{K(\delta h/\delta l)}{n} \tag{1.6}$$

where v is the seepage velocity and n is the effective porosity.

The dependence of permeability on the properties of the medium which include porosity, packing, void size distribution and void shape has led a number of investigators to attempt to relate permeability to the physical properties of porous media. Relations have been proposed by Hazen (1983), Slichter (1989), Terzaghi (1925), Smith (1933) and Ross (1945) among others. The Kozeny (1927)–Carman (1956) equation can be considered typical:

$$K = \frac{1}{C_T C_0 S^2} \frac{e^3}{1+e} \frac{\gamma_f}{\mu} \tag{1.7}$$

where C_T is a pore tortuosity factor, C_0 is a pore shape factor, S is the surface area of voids per unit volume of material or specific surface area, and e is the void ratio.

Other theoretical equations have suggested that a plot of K versus $e^2/(l+e)$ or K versus e^2 should be a straight line, while considerable experimental data indicates that e versus log K is frequently a straight line.

Although fundamental difficulties in the determination of the constants restrict the application of such formulae, they do serve to illustrate the relative importance of porosity, pore nature and fluid properties in regulating the rate of discharge.

1.2 PERMEABILITY OF DISCONTINUOUS MEDIA

The preceding expressions are only strictly applicable to steady-state, laminar flow through saturated, homogeneous, isotropic media. As such they have been found to lend themselves to much theoretical analysis concerning the flow of fluids through soils and other porous media. The application of these basic laws of flow to the problems of water percolation through rock masses is a more complex issue because of the highly jointed nature of most rock masses. Although it is possible to justify the assumption that some rocks *en masse* constitute natural porous media by virtue of their having a small fracture spacing in comparison with the dimensions of the rock under consideration, a more rigorous approach involves the analysis of flow through discontinuities in an effectively non-permeable rock material.

Discontinuities occur in all rock types and in a spectrum of scales from ångströms to kilometres. Micro-discontinuities such as intra-crystalline defects and fissures at crystal interfaces are responsible for the rock matrix permeabililty. Larger discontinuities such as joints, faults, large fractures, bedding planes, cleavages and shear fractures can, when interconnected, constitute important flow paths. Large-scale discontinuities such as regional faults and crush zones are potentially major water conduits, primarily because of their individual continuity.

As a first approach to the analysis of flow in fissured media, it is customary to approximate the discontinuity geometry to that of an equivalent simple structure, usually a regular system of plane, frequently horizontal, smooth fissures and then to describe the flow regime by analogy with laminar flow between parallel plates.

The magnitude of the hydraulic conductivity of a single discontinuity, parallel to its plane, is given by:

$$K_j = \frac{\gamma_f a^2}{12\mu} \tag{1.8}$$

where a is the effective aperture and K_j is the discontinuity conductivity.

This is expanded in Chapter 4. The hydraulic conductivity of a joint set is constant in every direction parallel to the set and its magnitude is given by:

$$K_s = \frac{a}{s} K_j + K_m \tag{1.9}$$

where K_s is the set conductivity, K_m is the rock matrix conductivity and s is the mean joint spacing.

In many circumstances the hydraulic conductivity of the rock matrix will be negligible in comparison with the term $(a/s)K_j$. Consequently, for discontinuity flow:

$$K_s = \frac{a}{s} K_j. \tag{1.10}$$

Where the rock itself is porous or poorly jointed, or where the joints are filled with material of low permeability, the above approximation will be invalid and the rock matrix permeability must be considered.

Although originally formulated for the analysis of porous media, the Kozeny–Carman equation has been extended to discontinuous media by analysing fractures as elongated pores. Fisher (1977) reported the equation to be of the form:

$$k = \frac{n^3 s^2}{180}. \tag{1.11}$$

Snow (1968a) utilized an idealized parallel plate model to derive a relation between total porosity and permeability. For two-dimensional flow in equally spaced orthogonal joint sets:

$$k = \frac{n^3 s^2}{96}. \tag{1.12}$$

For a cubical joint set the expression:

$$k = \frac{n^3 s^2}{162} \tag{1.13}$$

closely approximates the Kozeny–Carman equation.

These expressions are developed further in Chapter 4. It is important to remember that in the application of these expressions there are limitations. In particular a deterministic relation between rock mass permeability and porosity does not exist because these properties are not directly related to each other.

1.3 LABORATORY PERMEABILITY STUDIES – INTACT ROCK SPECIMENS

In the laboratory, permeability measurements are generally made on specimens of from 25 mm to 150 mm minimum dimension, which are typically intact. Two basic techniques are used. When the anticipated permeability is above about 1 microdarcy, equivalent to a hydraulic conductivity of approximately 10^{-11} m/s at normal temperature, a steady-state measurement is made in which the volume of fluid discharged through the sample is measured per unit time under a fixed pressure gradient.

For rocks of lower permeability, transient or slug tests – in which the decay of a pressure pulse applied suddenly to one end of a sample is monitored – have proved more suitable. Recent refinements of the transient method are thought to have lowered the minimum detection level to around 10^{-12} darcies (10^{-17} m/s).

The use of intact samples implies that laboratory values reflect the matrix permeability of the rock mass, micro-discontinuities constituting the principal flow paths. Other than in exceptional circumstances, where a system permeability is approximately equal to the rock matrix permeability, laboratory values will tend to represent the lower bound of *in situ* measurements.

Typical ranges of hydraulic conductivity measured in the laboratory for a number of generic rock types are summarized in Figure 1.1, together with comparable hydraulic conductivities from *in situ* measurements which are discussed below.

A number of aspects of reported laboratory values are worth noting:

1. Several orders of magnitude separate the most from the least permeable rocks of a particular geologic description. Considering the limits of detection of contemporary instruments this range could be even higher.
2. The greatest variation is observed for carbonate and volcanic rocks.
3. Permeability does not correlate directly with porosity, although broad relations have been established between the two parameters for individual formations and lithotypes (e.g. Archie, 1950; Reeves, 1977).
4. Even on a laboratory scale rocks show appreciable anisotropy. Shales and clays typically have ratios of horizontal to vertical permeability in the range 0.5 to 10, a representative value being 2 or 3 (Freeze and Cherry, 1979; Isherwood, 1979; and Mitchell, 1976). The direction of principal permeability is parallel to the preferred orientation of the clay particles.

   Studies of the directional permeability of sandstone cores by Greenkorn and Johnson (1964) gave ratios of maximum to minimum permeability of from 1 to 1.5. The major permeability axis was found to correlate with the strike of the formation bedding, while the minor permeability axis correlated with grain size. Zoback and Byerlee (1976) found that the permeability of Berea Sandstone parallel to the bedding was approximately four times that normal to the bedding.

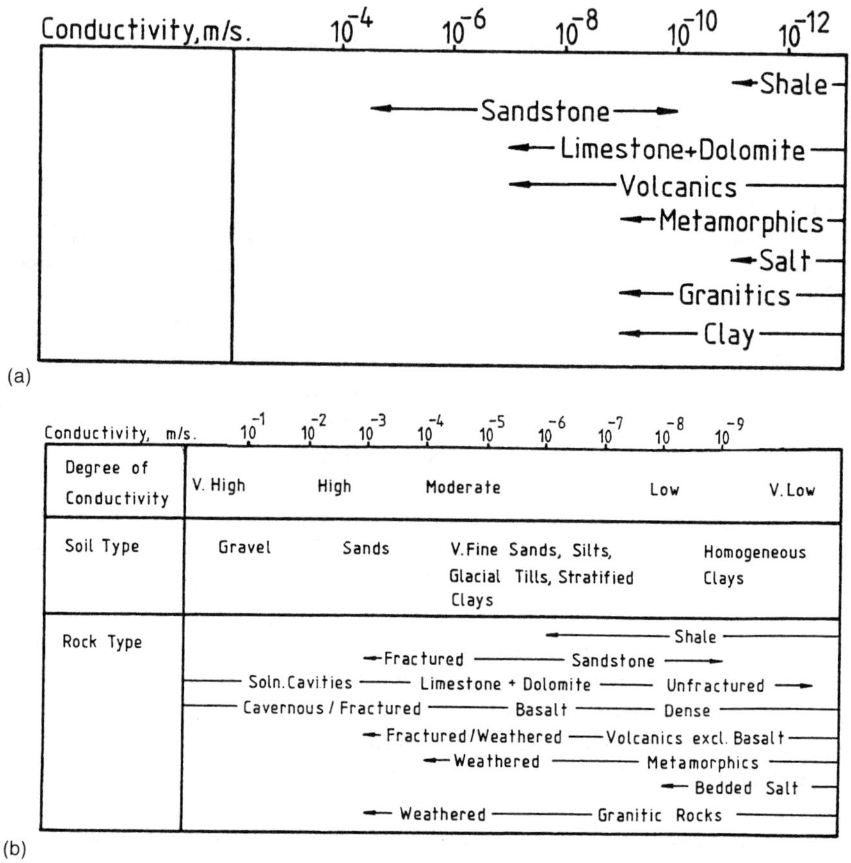

Figure 1.1 Representative values of hydraulic conductivity: (a) from laboratory measurements, (b) from *in situ* permeability measurements in rock masses (after Isherwood, 1979). Equivalent ranges of rock porosity are: shales 0.5–4.5%; sandstones up to 50%; limestones up to 10%; volcanics up to 1%; metamorphics up to 2.5%; granite 0.05–3%; salt up to 1%; clays up to 55%.

Samples of granite tested by Pratt *et al.* (1977) were found to have permeabilities parallel to the principal jointing approximately 2.5 times those normal to the jointing, a consequence of preferential orientation of micro-fractures within the rock.

In rocks such as slate and gneiss which exhibit significant foliation, it is not uncommon to find that permeability parallel to the foliation is considerably greater than that normal to the foliation.

Anisotropy of intact rock is primarily a function of the preferred orientation of mineral particles and micro-discontinuities. In nature the permeability of intact rock parallel to the bedding or stress field tends to exceed that perpendicular to it by a small factor.

5. During extended testing, measured values of permeability tend to decrease with time, changes of over three orders of magnitude having been recorded. Reynolds (1960) found that a period of from 5 to 10 days was generally required for conditions of steady-state flow through salt to be achieved, a magnitude supported by the results of Sutherland and Cave (1978), which can be interpreted as implying steady-state conditions at constant confining pressure after about 6 days. Explanations for this phenomenon include:

   (a) **Viscoelastic compaction:** Movement of deformed grain volumes to fill interstitial granular voids, thereby decreasing the void area capable of transmitting fluid.
   (b) **Surface tension:** A strong attraction between dry interfacial substrate molecules and fluid molecules gives rise to a high initial flow rate which decreases with time.
   (c) **Plugging effects:** Air or contaminants carried by the permeant cause pores to be plugged.
   (d) **Chemical reaction** between permeant and sample.

1.4 ROCK MASS STRUCTURE

Following deposition, sediments tend to show a gradual decline in primary porosity and permeability due to diagenetic processes of compaction, consolidation and cementation. By contrast, igneous rocks crystallize from the magma with minimum porosities and permeabilities. Hydrologically, however, subsequent alteration of these two basic rock types proceeds along basically similar lines. Tectonic and other environmental stresses are relieved as the various types of discontinuities which give rise to secondary porosity and permeability develop, while sub-surface circulation of fluids may enhance or reduce both porosity and permeability by processes of solution and precipitation. Consequently, in most rock formations, primary rock properties are relatively unimportant as far as the transmission of fluids is concerned, which tends to be principally a function of the comparatively high effective permeabilities of secondary discontinuities. Secondary porosities are, however, enhanced to a much lesser degree by discontinuity formation and in some younger and coarse grained sediments groundwater storage may remain an essentially intergranular property.

Intra-crystalline defects such as dislocations, intrusions and, on a slightly larger, scale, fissures at crystal interfaces, are common in almost all rock types. Analogue studies have indicated the probability of a comparatively high degree of interconnection between sets of microcracks and both individual microcracks and larger scale discontinuities.

Larger scale discontinuities such as joints, faults and other types of fracture from diverse origins frequently constitute pervasive potential water conduits. Stratigraphically engendered joints and partings may create flow paths parallel to the bedding. Cleavages or foliation derived from preferential orientation of platy or elongated minerals often give rise to planar, parallel conduits in metamorphic rocks. Extension fractures associated with cooling, folding, uplift and induced tension frequently constitute rough-walled conduits which may be partially or completely filled with hard or soft secondary deposits. Shear stresses can produce parallel minor faulting. Flow of cooling volcanic fluids may give rise to 'flow banding' or interconnected vugs aligned and extended in the direction of flow. Volumetric expansion associated with stress relief causes the close fissuring characteristic of granitic rocks seen in quarry exposures. Crystalline carbonate and granitic rocks are especially prone to fissuring by thermal stresses. Selective dissolution can result in major flow conduits, particularly in carbonate rocks or inhomogeneous deposits containing soluble horizons. Natural hydro-fractures may be totally or imperfectly sealed with minerals. Contacts between dissimilar formations may present fracture zones on both sides of the interface due to different mechanical properties.

Such fracture networks are often exceedingly complex but are not random and can thus be characterized by statistical methods which are discussed further in Chapter 2. Proposed statistical distribution functions are many, including negative exponential (Priest and Hudson, 1976), log-normal or gamma (Baecher *et al.*, 1977) and Weibull (Vasilev *et al.*, 1978). Far less is known concerning other characteristics of natural discontinuity features including apertures, infillings, spatial extents, relative surface roughness and mechanical properties which are not adequately sampled by conventional exploration techniques.

Finally, minor faults, regional faults and crush zones may traverse any rock type at any orientation. Their individual continuity makes them potentially major water conduits, although their exact nature, particularly the presence and characteristics of infilling materials, will exert a significant influence on their actual hydraulic properties.

The behaviour of groundwater as it percolates through a rock mass is thus a complex function of a large number of factors, some of which are sampled during *in situ* testing.

1.5 *IN SITU* INVESTIGATIONS

In situ values of permeability refer typically to measurements obtained by testing intervals of vertical boreholes; as such they reflect some average coefficient for an interval of strata adjacent to the borehole. Three principal techniques are utilized:

1. the Lugeon, packer or injection test in which the flow from a section of borehole is measured as a function of the applied pressure;
2. slug or pulse tests in which pressure recovery or decline is observed as a function of time following disruption of the pre-test equilibrium;
3. tracer tests in which the movement of chemically marked fluid is observed.

Such test techniques (Ziegler, 1976) aim to induce primarily radial flow to or from a test site and thus tend to preferentially sample the horizontal permeability of a rock mass.

During the following discussion, values of permeability are utilized as documented, no attempt having been made to assign probable errors. Because of fundamental difficulties in the interpretation and analysis of data obtained from *in situ* permeability tests, the presence of systematic errors is inevitable. The general range of the values of *in situ* permeability observed for a number of generic rock types is summarized in Figure 1.1(b).

A number of aspects of reported *in situ* values are worth noting:

1. The range of *in situ* values is significantly greater than the range of laboratory values, spanning over 12 orders of magnitude. At a particular site a variation of four to six orders of magnitude is typical, while permeability variations of up to four orders of magnitude have been observed on individual sites at the same depth horizon. This is hardly surprising when the dependence of permeability on fracture aperture and frequency is considered; a major active discontinuity can easily increase the average permeability of a test section by orders of magnitude over adjacent sections.

   Although a general relation is often apparent when fracture frequency and measured permeability are plotted against depth, the correlation is by no means absolute – permeability is not a simple function of total fracture frequency. The correlation between the two parameters can, however, be improved if the number of active fissures – those which, by virtue of staining or other characteristics, give some indication of being water conducting – is considered (Reeves, 1977). McCrae (1982), for example, estimated that only some 20% of discontinuities encountered during construction of the Muna highway tunnels, Saudi Arabia, were potential water conduits, of which only about 25% had positive evidence of being so.
2. Rock mass permeability is often significantly anisotropic. The ratio k_h/k_v is thought to vary over several orders of magnitude, perhaps from 10^{-2} in fractured rocks whose joints are primarily vertical to 10^3 in intact layered sedimentary strata. For analytical purposes, values of $k_h/k_v = 1$ or 10 are commonly used.

   Tests conducted in salt formations suggest that the k_h/k_v ratio generally approximates to unity, although at the WIPP (Waste Isolation Pilot Project) site in bedded salt, the horizontal stratum permeability was estimated to be from 10^2 to 10^3 greater than the vertical permeability based on a measured *in situ* horizontal hydraulic conductivity of 5×10^{-11} m/s in a section incorporating a more permeable interbed (Griswold, 1977).

   Rocha and Francis (1977) developed a method for estimating the permeability tensor from the characteristics of integral samples by applying a correction factor derived from the analysis of *in situ* pressure tests. Their results for a basaltic rock mass showed a systematic near-vertical direction of minimum principal permeability and an almost isotropic horizontal permeability with a magnitude considerably greater than k_v. The coefficient of anisotropy lay, as a rule, in the range 5 to 50 but two values exceeding 1000 were also obtained. Interestingly, sedimentary interbeds in basaltic formations may constitute either aquifers or aquicludes, depending on whether they are more or less permeable than the basalt itself.
3. It is possible to infer values of permeability from various large-scale phenomena, principally those thought to be controlled by hydraulic diffusion. These include

hydrothermal intrusions, earthquake-associated events, maintenance of pore fluid pressures, convection within the oceanic crust and aquifer contamination. In a transient flow regime, as with chemical diffusion or heat flow, a basic theoretical proportionality exists, for a moving pressure front, between time and the distance squared such that:

$$t \propto \frac{L^2}{\alpha} \tag{1.14}$$

where t is time, L is distance and α is hydraulic diffusivity. Here

$$\alpha = \frac{k}{n\mu B} \tag{1.15}$$

where k is permeability, n is porosity, μ is fluid viscosity and B is the fluid compressibility.

Given a phenomenon due to transient flow, t and L can be found and the permeability estimated from a knowledge of rock porosity and fluid properties. Many inferred permeabilities reflect values averaged over distances of kilometres and the volumes of rock represented are, therefore, much greater than those sampled by either laboratory or typical *in situ* tests. Permeabilities inferred in various ways are plotted in Figure 1.2. The close agreement of many of the values inferred from different phenomena over the range $10^{-1} - 10^{-4}$ darcies is quite striking. Not surprisingly, values inferred for joints and mining-induced fractures were considerably higher, while permeabilities inferred from the maintenance of anomalous fluid pressures were much lower.

A comparison of laboratory, *in situ* and inferred permeabilities suggests a number of important, though of necessity broad-based, conclusions.

1. For rocks such as mudstones and shales which have a significant clay content, laboratory and *in situ* values may agree to within an order of magnitude. Some intervals of test wells in other rock types, including crystalline rocks, may show a similar agreement in the absence of conducting discontinuities.
2. Conducting fractures may increase *in situ* permeabilities by three or more orders of magnitude over laboratory values.
3. Permeabilities inferred for argillaceous rocks may be within an order of magnitude of values given by *in situ* tests. For crystalline rocks most inferred permeabilities are of the same order as the more permeable zones *in situ*, the average permeability being from 1 to 10 mdarcies. This suggests an important conclusion: that pumping tests can sample volumes of rock representative of much larger volumes.
4. The average permeability of crystalline rocks noted above corresponds to that of many intact sandstones and limestones.

A question fundamental to a meaningful estimation of effective rock mass permeability concerns the dimensions of the sample which must be tested in order to obtain a value representative of the rock mass. As sample volume increases, it

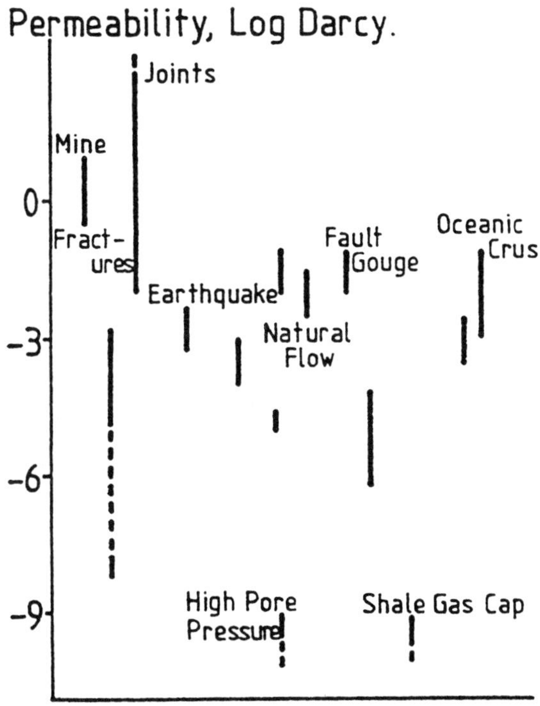

Figure 1.2 Permeabilities of structural features inferred from large-scale phenomena (after Brace, 1980).

would be expected that measured permeabilities will initially show considerable fluctuation, but will eventually tend to an approximate constant value as a sample size becomes representative and reaches the so-called Representative Elementary Volume (REV). This is considered further in Chapter 5. For volumes smaller than the REV, discrete fracture effects influence measured permeabilities as the test volume is increased by small increments. For volumes greater than or equal to the REV discrete fracture effects are no longer discernible and measured permeabilities are, therefore, representative of a much greater volume of the rock mass. Although it is not, as yet, possible to calculate the value of REV directly, packer tests which utilize an increasing scale of packer spacings offer some promise for its determination *in situ*. In this context the correlation noted in item 3 above between the permeability of the more permeable zones in test wells, which sample at most a few thousands of cubic metres of rock, and values of permeability inferred from phenomena which affect cubic kilometres of the earth's crust, is worthy of further investigation.

1.6 INFLUENCE OF GEOSTATIC STRESS ON ROCK PERMEABILITY

Theoretically and in the laboratory, it can be shown (Chapter 4) that the stress applied to an intact or fractured rock sample can have a significant influence on its effective permeability. Changes of over three orders of magnitude have been recorded at comparatively low stress levels; typically, permeability appears asymptotically to approach a minimum value with increasing normal stress. The question then arises as to the effect of geostatic stresses on the permeability of discontinuities within a rock mass. More specifically, does rock mass permeability vary systematically as a function of depth?

In general, the vertical component of virgin geostatic stress has a value which approximates that attributable to the weight of overburden and to which the equation:

$$\sigma_v = 0.027Z \tag{1.16}$$

where σ_v is the vertical geostatic stress in MN/m^2 and Z is the depth in metres, is a reasonable approximation.

Departures from this relation occur in areas of rapid erosion, areas of uneven topography and in proximity to intrusions of rock with different mechanical properties.

Measurements of the complete state of geostatic stress have been performed in many rock at different depths across the world. A compilation of these measurements, made by Brown and Hoek (1978), is summarized in Figure 1.3(a). The horizontal component of geostatic stress appears to range from about one-third to three times that of the vertical component. Relatively high values of horizontal stress tend to be comparatively shallow phenomena, possibly associated with the effects of denudation.

Jamison and Cook (1979) performed a similar compilation using data from tests where the three principal components of stress had been determined and where one principal stress was essentially vertical, as in Figure 1.3(b). With the exception of some very high values of the average horizontal stress near the surface, Figures 1.3(a) and (b) appear similar. Close inspection of these data reveals that they can be divided into three groups, depending on the value of the vertical stress compared with those of the other two principal stresses. Anderson (1951) shows that three types of faulting are possible according to the relative magnitudes of the principal stresses:

1. thrust faulting where the minimum principal stress is vertical;
2. strike-slip faulting where the intermediate principal stress is vertical;
3. normal faulting where the maximum principal stress is vertical.

These three groupings are identified in Figure 1.3(b). As would be expected, normal faulting stress conditions are those with relatively low values of the average horizontal stress, thrust faulting stress conditions are those with relatively high values of average horizontal stress, and strike-slip faulting stress conditions are those with intermediate values of average horizontal stress.

A majority of geostatic stress field measurements have been performed in relatively competent rocks such as quartzite, granite and Pre-Cambrian rocks. In softer rocks, it is commonly assumed that variation of the horizontal stress component will not be so marked due to long-term creep and stress redistribution. If permeability does vary as a function of geostatic stress, these observations of actual geostatic stress configurations have important implications for the permeabilities of rock masses.

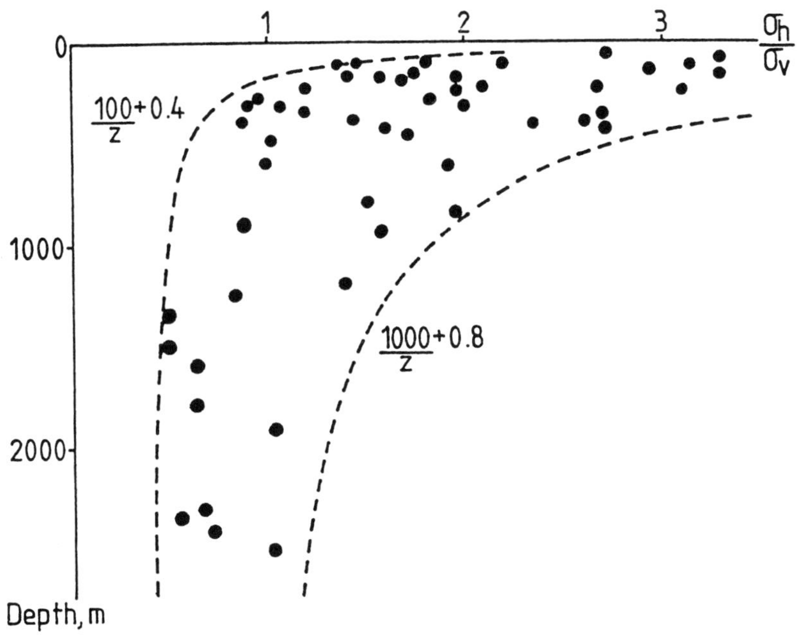

(a)

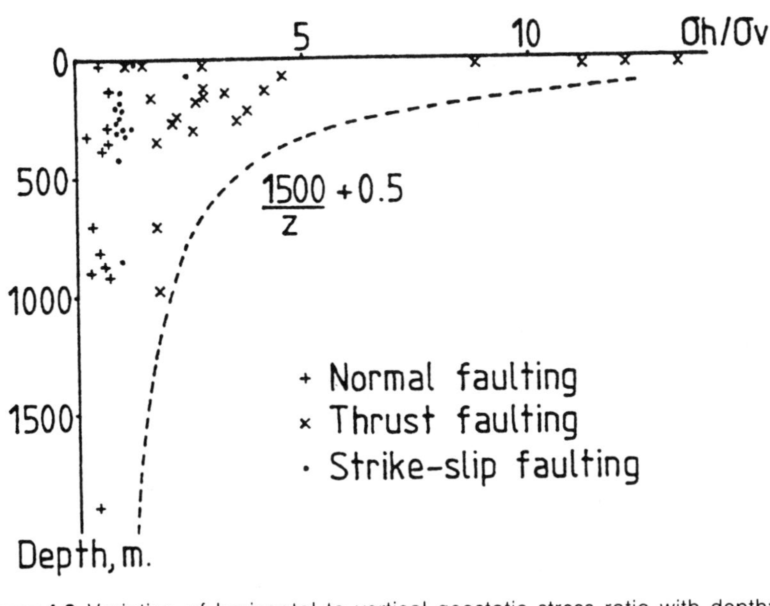

(b)

Figure 1.3 Variation of horizontal to vertical geostatic stress ratio with depth: (a) data collected by Brown and Hoek (1978); (b) data collected by Jamison and Cook (1979) with interpretation of faulting conditions.

Snow (1965, 1968a, b, c, 1970), Lewis and Burgy (1964), Uhl and Sharma (1978), Louis (1974), Davis and Turk (1964), Carlsson and Olsson (1977a,b) and others concur that rock mass permeability and fracture porosity decrease with depth. On the basis of hydrological data obtained from site investigations, a number of them have proposed and utilized empirical functions relating horizontal strata permeability, as measured by drill stem tests, to depth.

Snow observed that, although the permeabilities at similar depths at a number of dam sites varied, all seemed to follow a logarithmically declining trend with depth. On the basis of these data he suggested that the permeability of a fractured, crystalline rock mass could be estimated by plotting the mean permeability of each depth zone against depth and constructing a line through the points by means of a regression analysis. He concluded that permeability varies with depth according to the equation:

$$\log k = -8.9 - 1.67 \log Z \tag{1.17}$$

where k is the intrinsic permeability in ft^2 and Z is the depth below the surface of the rock mass in feet (1 ft = 0.3048 m).

Louis (1974) studied the influence of overburden stress on the hydraulic conductivity measured at various depths in boreholes in fractured formations. After an examination of numerous test results, he concluded that the empirical law which best translated the phenomenon was:

$$K = K_0 e^{-\alpha\sigma} \tag{1.18}$$

where K_0 is the surface hydraulic conductivity and σ is the overburden stress, $\approx \gamma Z$.

On the Grand Maison dam site, he observed that the coefficient $A = \alpha\gamma$ varied between 7.8 and $3.4 \times 10^{-3}\,m^{-1}$ according to the position of the test site and K_0 between 10^{-7} and 10^{-6} m/s.

Carlsson and Olsson (1977) performed a regression analysis on hydraulic conductivity/depth data from a number of sites in Sweden and found that two equations had comparatively high indices of determinations:

$$K = 10^{-(0.02Z + 5.6)} \tag{1.19}$$

$$K = 10^{-(2.5 \log Z + 2.5)} \tag{1.20}$$

where K has units of m/s and Z has units of m.

The indices of determination for the two equations were 0.40 and 0.57 respectively which, they suggested, was a comparatively high correlation for heterogeneous bedrock. Snow's equation in an equivalent form is:

$$K = 10^{-(1.6 \log Z + 4)} \tag{1.21}$$

Using values of permeability obtained from measurements in Sweden, a mean curve for the variation of horizontal hydraulic conductivity with depth has been estimated and fitted with a third-degree polynomial by Burgess (1977):

$$\log K = 5.57 + 0.352 \log Z - 0.978(\log Z)^2 + 0.167(\log Z)^3 \tag{1.22}$$

where K has units of m/s and Z has units of m.

Sources of data included water abstraction wells, drill holes in the Forsmark area and a deep drill hole at Oskarshamn. This equation has been transformed into one relating stress and hydraulic conductivity using the result $\sigma = \gamma Z$:

$$\log K = -5.57 + 0.352\left(\log\frac{\sigma}{\gamma}\right) - 0.978\left(\log\frac{\sigma}{\gamma}\right)^2 + 0.167\left(\log\frac{\sigma}{\gamma}\right)^3. \qquad (1.23)$$

Some disagreement has developed among authors who, while agreeing that permeability does decrease with depth, disagree as to whether the cause is increasing fracture spacings or decreasing fracture aperture. This is considered further in Chapter 4.

1.7 RELATIONS BETWEEN DISCONTINUITY GEOMETRY AND DEPTH

On the basis of the results of *in situ* permeability tests and fracture frequency determinations, Snow (1968a) observed that discontinuity spacings generally increase with depth while fracture apertures generally decrease. He found (Figure 1.4(a)) that effective discontinuity openings in the gneisses of the Front Range, Colorado, ranged from 75 to 400 μm in the upper 10 m of rock mass, declining to 40 to 100 μm at depths greater than about 30 m. Average fracture spacings (Figure 1.4(b)) ranged from 0.15 to 1.2 m near the surface, increasing to 1.5 to 4.5 m immediately below the zone of surface weathering and to 5 to 8 m at depths greater than 50 m.

Corresponding to the observed changes of fracture spacing and aperture with depth is a decline in effective fracture porosity. Snow (1968a) observed that the fracture porosity of the metamorphic rocks he investigated declined from 0.05% near the surface to approximately 0.0005% at a depth of 122 m and generalized that effective discontinuity porosity decreases approximately tenfold per 60 m depth. Snow's values of effective porosity are approximately three orders of magnitude smaller than the porosity estimates of Davis and Turk (1964), Shehata (1971) and Stewart (1962) for similar rock types, suggesting a significant difference between total and effective porosities. Effective porosity is a measure of the void space which actively contributes to flow through a porous medium and is generally a fraction of the total medium porosity. Although the effective porosities of rocks such as sandstones probably lie in the range 6–20% (Isherwood, 1979) and thus may constitute a significant proportion of total porosities, the effective porosities of fine-grained rocks such as siltstones and shales may be considerably smaller. Effective porosities of fractured rocks with low interstitial porosities may be as low as 10^{-4} (Freeze and Cherry, 1979) or even less (Snow, 1968a). These estimates are consistent with other observations of the hydraulic characteristics of rock mass discontinuities.

Snow's (1968a) estimate of the effective porosity of the Idaho Springs formation at the Rocky Mountain Arsenal well, about 0.04% at the surface, decreasing to 0.001% at a depth of 100 m, was substantiated by reservoir calculations of Van Poolen and Hoover (1970). He also noted that fracture porosity depends more on fracture spacings than on apertures and orientations, and that neither aperture width nor fracture

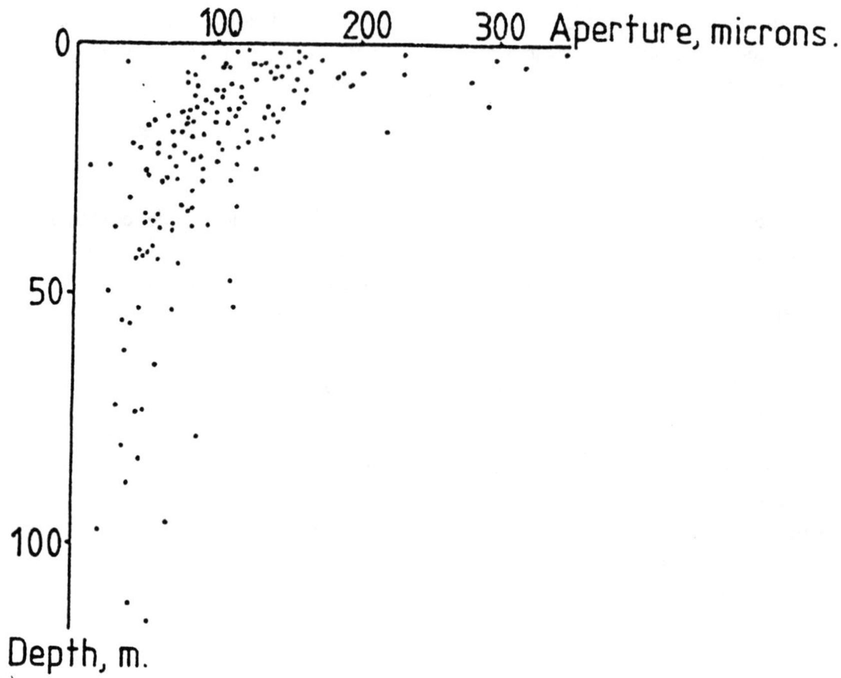

100
200
300
Aperture, microns.
0
50
100
Depth, m.

(a)

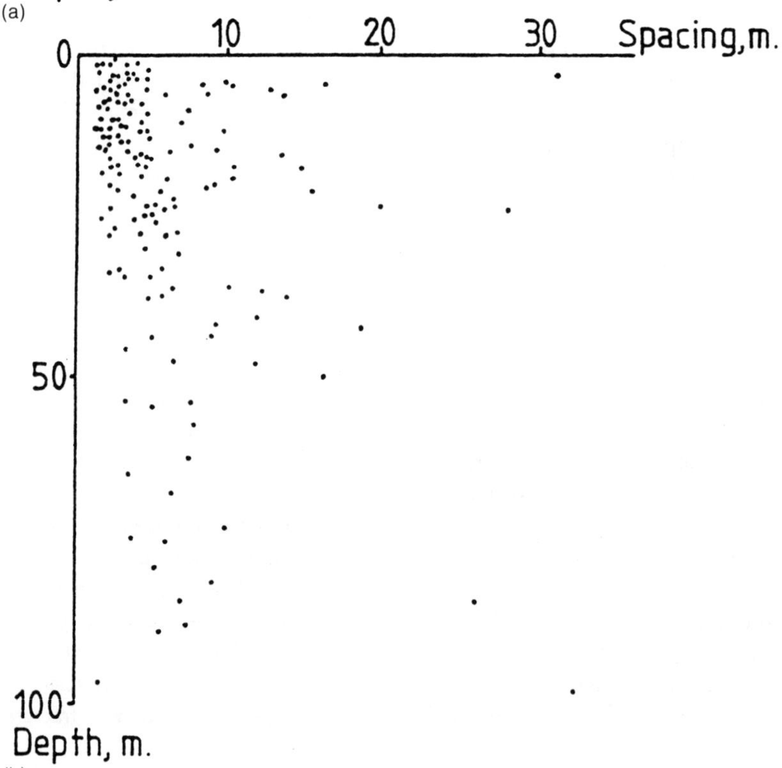

10
20
30
Spacing, m.
0
50
100
Depth, m.

(b)

spacing differs notably from one igneous/metamorphic rock type to another, nor do the associated porosities and permeabilities.

Webster *et al.* (1970) found two types of discontinuity common in the crystalline rock at the Savannah River Plant, South Carolina. The first type pervades the entire rock mass but transmits very little water. The second type is restricted to definite zones and transmits substantial quantities of water. The hydraulic conductivity of the rock mass was 5×10^{-9} m/s and the calculated fracture porosity 0.08%. The effective fracture porosity of rock containing only the first type of fracture was thought to be 0.01% or less.

Cloos (Kendorski and Mahtab, 1976) identified an orthogonal system of joints characteristic of almost all granitic platons:

1. Q-fractures resulting from extensive fracturing parallel to the principal direction of stress;
2. S-fractures resulting from failure along anisotropies resulting from flow;
3. L-fractures resulting from variations in vertical loading or rebound.

Additional fractures may be superimposed due to later tectonic activity.

Price (1966) confirmed these basic concepts but concluded that primary joints had little hydraulic significance. He proposed that brittle fractures resulting from unloading and uplift were of the most significance to groundwater flow. Raven and Gale (1976), Kendorski and Mahtab (1976), Manev and Avramova-Tacheva (1970), and others, however, found that below a surface zone where fracture frequency is increased as a result of weathering and other stress relief effects, fracture intensity remains approximately constant with depth, as in Figure 1.5.

Decreasing permeability with depth was, therefore, ascribed to decreasing fracture apertures. Conversely, investigators working on the Stripa Project in Sweden concluded that discontinuities with significant apertures persist at depth but their spacing increases (ONWI, 1978).

The discussion is inconclusive; although average fracture spacing does seem to increase with depth, no systematic variation is apparent over the range of depths normally encountered during civil and mining engineering operations. Considerable confusion may, however, have been introduced into the debate by consideration of total, rather than conducting, discontinuity frequencies. There is a substantial amount of evidence that **average** discontinuity apertures decrease with depth, although conducting discontinuities with significant apertures also appear to persist to depth.

The apparent persistence of discontinuities with significant openings at depth is consistent with the findings of laboratory studies. In the laboratory, normal stresses of over 20 MN/m^2, corresponding to the effective vertical stress at a depth of over 1000 m, have failed to affect either hydraulic or acoustic closure of comparatively smooth discontinuities, while tests conducted in the field led Snow (1965) to observe that, once formed and opened, a fracture is never completely hydraulically closed.

◄

Figure 1.4 Variation of (a) discontinuity aperture, and (b) discontinuity spacing with depth (after Snow, 1968a).

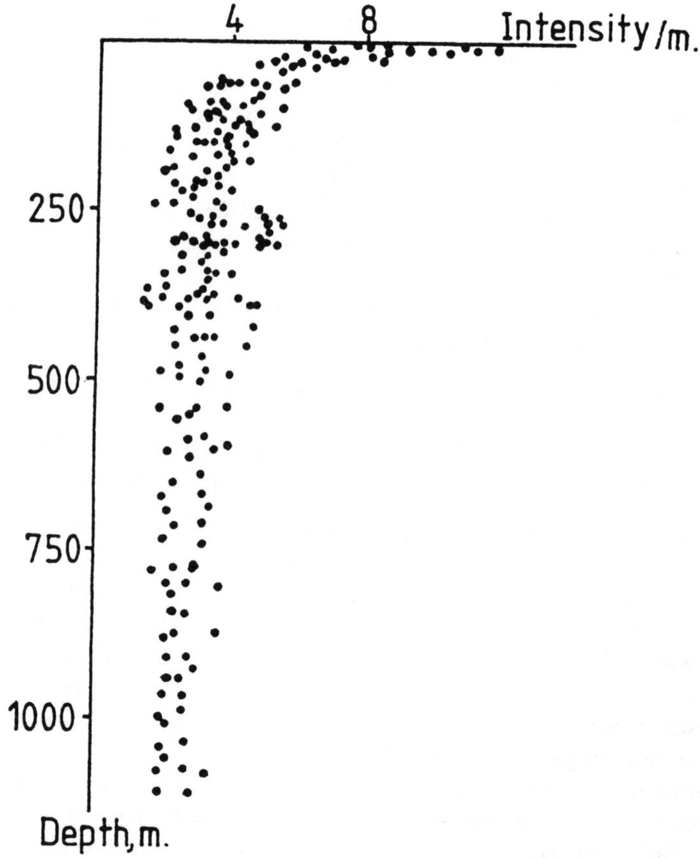

Figure 1.5 Variation of discontinuity intensity with depth (after Manev and Avramova-Tacheva, 1970).

Although individually extensive vertical joints, but not faults and other singular features, are comparatively rare on account of their tendency to terminate against horizontal discontinuities, they do constitute important components of flow paths between the surface and underground excavations.

The depth to which vertical joints remain open *in situ* is controversial. King (1948) observed joints which extended vertically as much as 330 m, while Chatterji (1966) found that at depths below 60 m vertical joints 'tightened up', but also observed that they tended to be more water bearing than horizontal discontinuities. On the basis of laboratory tests, Westerman *et al.* (1982) concluded that vertical discontinuities with low surface roughnesses might be expected to remain acoustically open to depths of at least 3 km.

1.8 MAXIMUM DEPTH FOR EXTENSION JOINTS

Extension joints form perpendicular to the minor principal stress, σ_3, under several different conditions. The total value of σ_3 may be tensile, in which case the joints are true tension joints. Alternatively, the total value of σ_3 may be compressive, but tensile stresses may exist locally on a granular scale; or the total value of σ_3 may be compressive, but the pore pressure may be sufficient to make the principal effective stress $(\sigma_3 - u)$ tensile.

In principle, there is a limiting depth for true tension joints which can be estimated from the Mohr envelope for a given rock type, as shown in Figure 1.6(a). Assuming the major principal stress, σ_1, to be vertical and due entirely to the weight of overlying

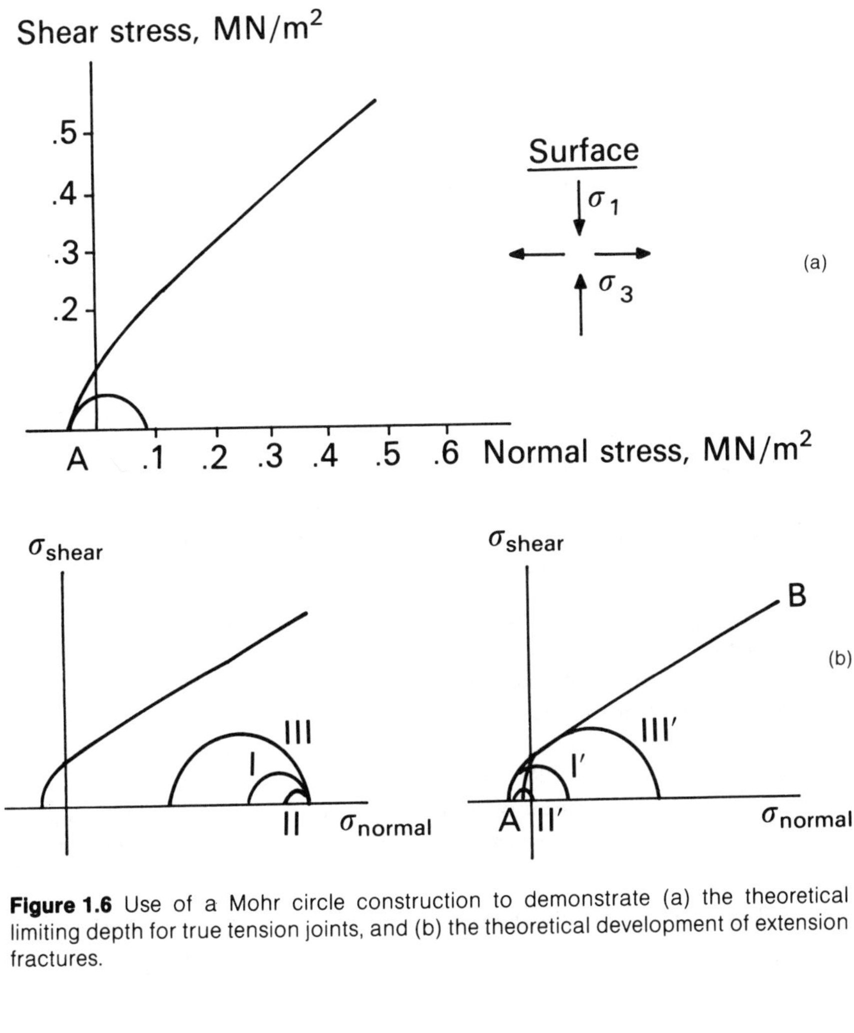

Figure 1.6 Use of a Mohr circle construction to demonstrate (a) the theoretical limiting depth for true tension joints, and (b) the theoretical development of extension fractures.

rocks, the state of stress just prior to tension jointing is represented by a circle that passes through point A (the uniaxial tensile strength) but is otherwise wholly within the stable region. If the largest Mohr circle that satisfies these requirements is constructed, its σ_1 value will correspond to the maximum depth at which tension jointing can occur. The difficulty with this procedure in practice is that it will tend to give excessive maximum depths because the Mohr envelope is based on short-term laboratory testing and does not express the time-dependent weakening effects expected in natural environments. Thus, while Figure 1.6(a) suggests a maximum depth of about 2.5 km for tension jointing, Griggs and Handin (1960, p. 351) suggest that macroscopic tensile stresses are inconceivable, even in stronger rocks, at depths below a few hundred metres.

Extension joints associated with compressive values of the total stress σ_3 can form to indefinitely greater depths than true tension joints, as long as pore pressure is high enough and the stress difference $(\sigma_1 - \sigma_3)$ is in the right range. This is illustrated in Figure 1.6(b) in which circles I, II and III represent the total stresses and circles I′, II′ and III′ represent the corresponding effective stresses. In the situation represented by I and I′, the pore pressure $u = 0.8\sigma_1$ and an extension fracture develops because the circle I′ touches the Mohr envelope at point A only. In situation II′, no fracture forms even at $u = 1.0\sigma_1$, because the difference $(\sigma_1 - \sigma_3)$ is less than the tensile strength. In situation III′, the stress difference $(\sigma_1 - \sigma_3)$ is greater than the tensile strength, but a shear fracture forms instead of an extension fracture, because the stress difference $(\sigma_1 - \sigma_3)$ is so great that the circle impinges on the Mohr envelope at a point between A and B.

1.9 EFFECT OF DEPTH ON PERMEABILITY

Work by Brace (1980) contradicts to some degree the generally accepted concept, illustrated in Chapter 4, of systematically declining permeability with depth. He assembled *in situ* permeability measurements for crystalline rocks from a number of locations, divided them into three depth zones – the surface to 50 m, 50 to 550 m and below 500 m – and presented the data in the form of a histogram (Figure 1.7). From these data he drew the following conclusions:

1. Permeability varies by over four orders of magnitude at a particular site within the same depth interval.
2. At nearly every site some portion of the rock sampled by the drill holes has a permeability of 1 m darcy to 100 m darcies. This relatively permeable zone may be as deep as 1.8 km or even 3.3 km. Similar zones were noted in a survey of 25 Canadian mines to 1.7 km (Raven and Gale, 1977).
3. There is a hint of a decrease of permeability with depth but hardly enough to justify the empirical laws suggested. Particularly for the values at depths greater than 500 m, no systematic variation with depth is evident.

If the data upon which empirical depth/permeability relations have been established are re-examined, it is apparent that a significant proportion were derived from

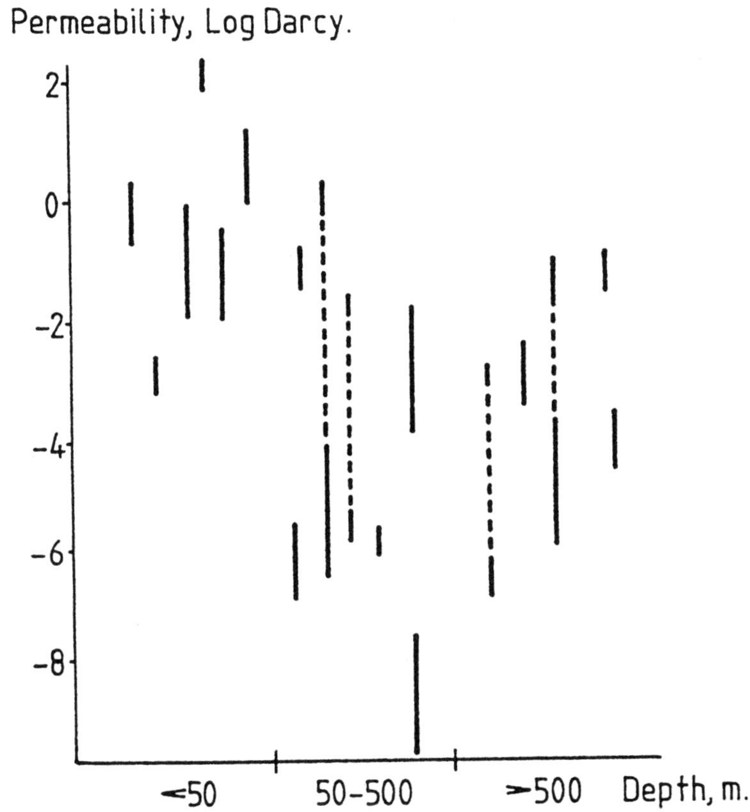

Figure 1.7 Variation of crystalline rock permeability with depth (after Brace, 1980).

comparatively shallow well and borehole tests, and therefore relate to depths where a zone of increased porosity and permeability might be expected as a result of surface phenomena. The scatter of data is often considerable and extrapolation to depth is frequently on the basis of a small number of data points. If these relate to rock below the zone of major surface effects, they may be more representative of the rock mass permeability at depth, rather than indicative of a systematically declining permeability with depth. The commonly observed increased 'incidence of zero-take tests' with depth may, therefore, also be evidence that zones of significant permeability persist to at least engineering depths. Additionally, no physical significance has, as yet, been ascribed to any of the depth/permeability functions presented.

In Figure 1.8, information concerning the *in situ* variation of hydraulic conductivity with depth is plotted for crystalline and sedimentary rocks. Although there is a hint of a decrease in permeability with depth, there is hardly enough evidence to justify the depth/permeability functions presented on a generic scale. A zone of comparatively high permeability does seem to be associated with the ground surface. Data from the sedimentary rocks, primarily 'shales' in this instance, may be somewhat misleading

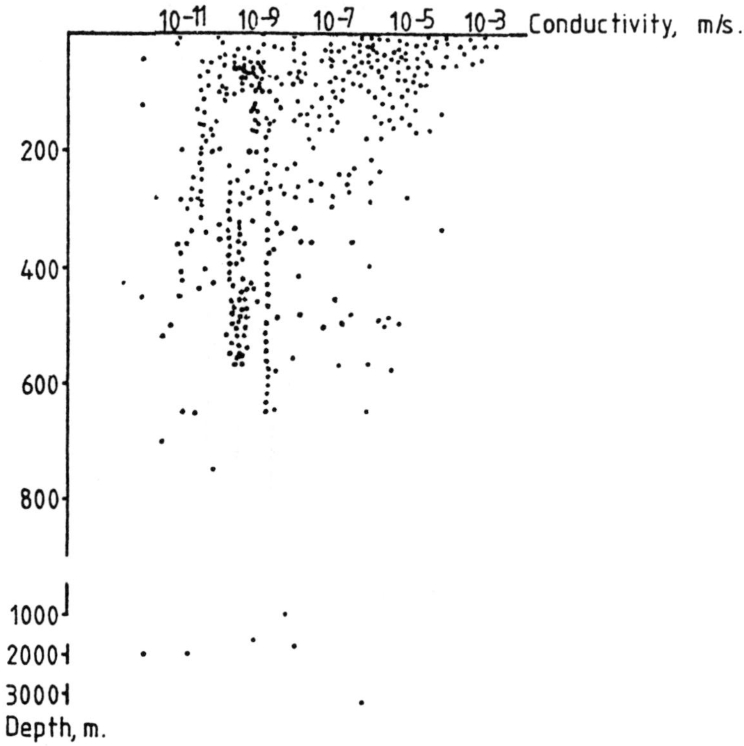
10-11 10-9 10-7 10-5 10-3 Conductivity, m/s.
200
400
600
800
1000
2000
3000
Depth, m.

(a)

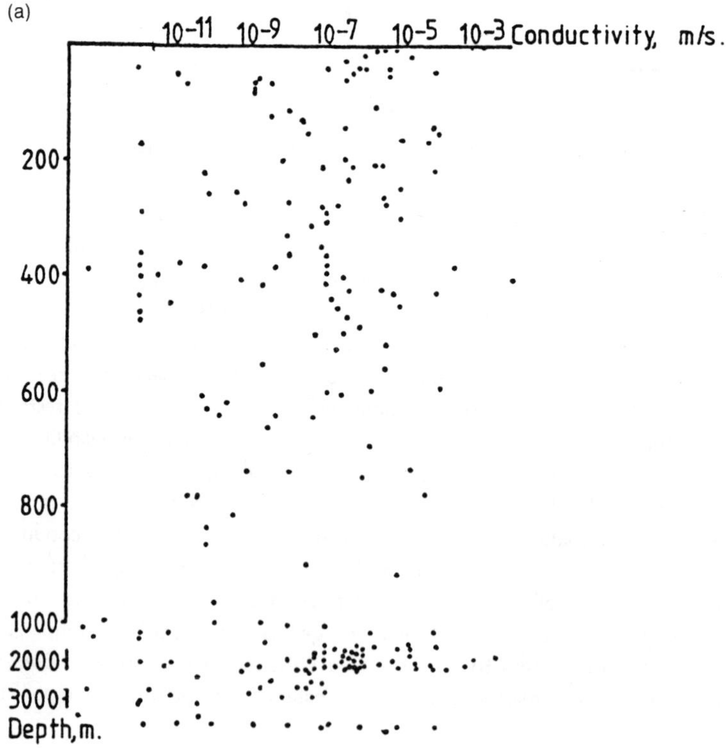
10-11 10-9 10-7 10-5 10-3 Conductivity, m/s.
200
400
600
800
1000
2000
3000
Depth, m.

(b)

in that permeability tests tend to be conducted at the more permeable horizons in oil and gas wells; however, the basic conclusion, that permeable horizons persist to depth, remains.

Although asymptotic values of effective fracture permeability with stress (depth) have been suggested, their general validity is uncertain. Given the wide 'limits' within which rock mass permeability may vary, it is particularly critical to define on a site-specific basis the hydraulic charcteristics of a formation.

1.10 IMPOSED ROCK FRACTURES AND PERMEABILITY PERTURBATIONS

In an introductory discussion of permeability it is important to discuss the role of fractures imposed on rock masses. Thus far discontinuities or fractures have been assumed to be natural, forming part of the rock mass and created by solidification, diagenesis, cooling, weathering and other natural processes. However, a major interest is the effect of fractures created by excavation and the effect they have on the near-field permeability around the excavation. This can best be approached by a simple study of the mechanics of fracture.

The mechanics of deformation of rocks and other brittle materials are well known. On a microscale, tensile stresses, induced at the tips of microcracks in the rock during deviatoric loading in compression, activate and nucleate these cracks beyond a stress threshold equal to about half the compressive strength, dilating the rock and increasing permeability. This is illustrated in Figure 1.9 where axial stress to failure of a typical axially loaded laboratory specimen is plotted against volumetric strain, axial strain and permeability.

The deformation characteristics are most clearly illustrated in the stress–volumetric strain curve. During initial elastic deformation this is linear and related to the stress–axial strain curve. Both represent compression (a compression positive plot is used) and the volumetric strain comprising the sum of axial compression and lateral expansion is theoretically equal to one-third of the axial strain. At about half the failure stress, microcrack nucleation commences in a direction parallel to the major principal (in this case axial) stress and increases at a progressive rate with increasing stress to failure or rupture. Towards rupture the lateral dilation increases rapidly, exceeding the linear compression, and is accompanied by a rapid increase in both volume and permeability.

Confining pressure has a pronounced effect on the stress–strain properties of rocks. Figure 1.10 illustrates this effect in tests on granite specimens. Increasing the confining pressure to 48 MPa 'strengthens' the rock and results, or would result, in an eventual transition from strain softening to strain hardening behaviour. Other

◄

Figure 1.8 Variation of hydraulic conductivity with depth in (a) igneous rocks, and (b) sedimentary rocks (from data collected by Garritty, 1983).

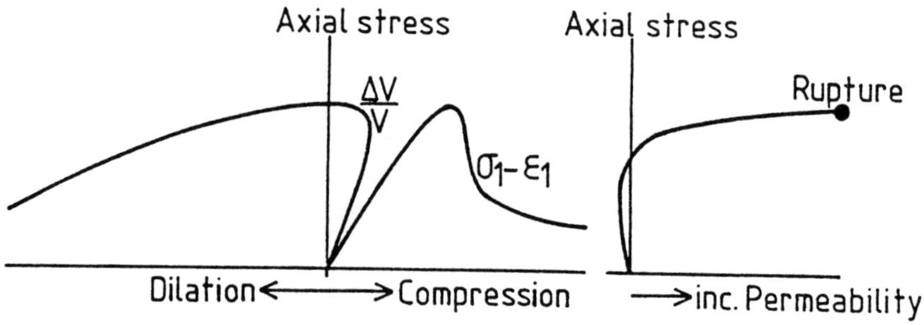

Figure 1.9 Typical stress, volumetric strain and permeability during testing in axial compression of a triaxially confined cylindrical rock specimen.

similar curves are illustrated in Farmer (1983). There is an increase in the axial stress to failure and a shallower post-failure curve with increasing confining pressure.

The volumetric strain against axial strain curves show a systematic change with confining pressure. All the curves have the same general form as Figure 1.9, a small volume decrease occurring at low strains, the slope of the curve changing, often rapidly, in the yielding region to a final near-linear volume increase after yielding. In all rocks uniaxially tested specimens show very large volumetric expansions with increasing axial strain, typically of the order of 10–20% axial strain. The effect of confining pressure on volumetric expansion is very marked, with a large reduction in dilation for a small increase in confining pressure. At all pressures below the threshold level, axial deformation leads first to compression of the specimen, corresponding with the linear elastic region of the axial stress–strain curves. The onset of dilation occurs during the linear part of this curve, due to microfracturing within the specimen. Rapid dilation coinciding with specimen failure is followed by almost linear volumetric expansion as frictional sliding takes place.

The greatest increases in permeability associated with yield are therefore to be expected in rocks transected by tensile or shear failure planes in a direction parallel to those planes. Even a modest confining pressure will tend to exert a significant influence on the final permeability of the rock by inhibiting dilation. The increases in permeability associated with primarily ductile deformation will tend to be significantly lower than those resulting from brittle failure, although 'final' permeability will still be appreciably greater than that of the undisturbed rock.

The effect of confinement on permeability is illustrated in Figure 1.11. Brace (1978) investigated the effects of sample failure on the permeability of Westerly Granite, a very brittle rock in confined compression. After an initial decrease, the permeability of samples parallel to the maximum compressive stress increased by a factor of from three to four at 80% of fracture stress.

Mordecai and Morris (1970) studied changes in the permeability of Darley Dale Sandstone, a quite weak sandstone, prior to failure at confining pressures up to 41 MPa (Figure 1.11). At a confining pressure of 7 MPa the behaviour was similar to that observed by Brace, although the absolute increases in permeability were much

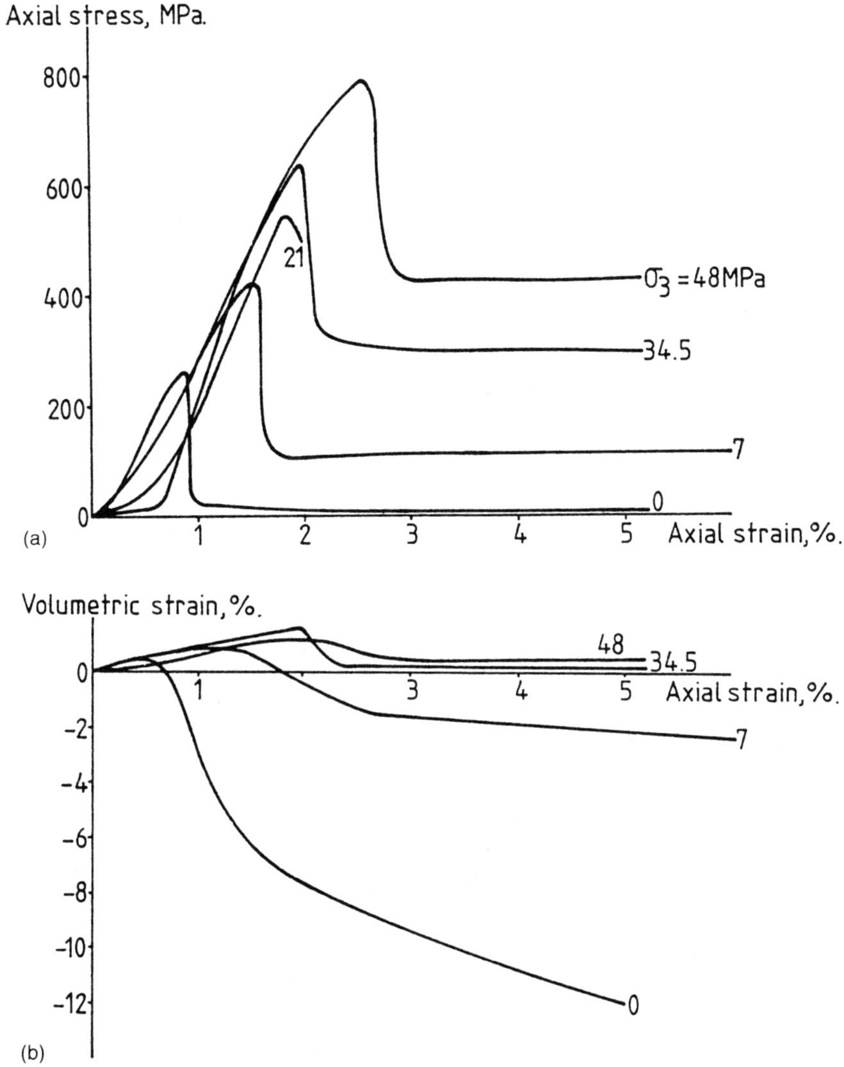

Figure 1.10 Axial stress–axial strain (a) and axial–volumetric strain (b) curves for 100 mm diameter by 200 mm long specimens of granite tested at confining pressures of 0, 7, 21, 34.5, and 48 MPa, to axial strains of about 5%.

lower. With increasing confining pressure the increase in permeability decreased, and at 41 MPa confining pressure the sample had actually undergone a 5% reduction in permeability immediately prior to failure.

Tests on granite sand by Zoback and Byerlee (1976) showed that permeability dropped by about one to two orders of magnitude up to peak stress, and although there appeared to be a small amount of dilation prior to peak stress, permeability

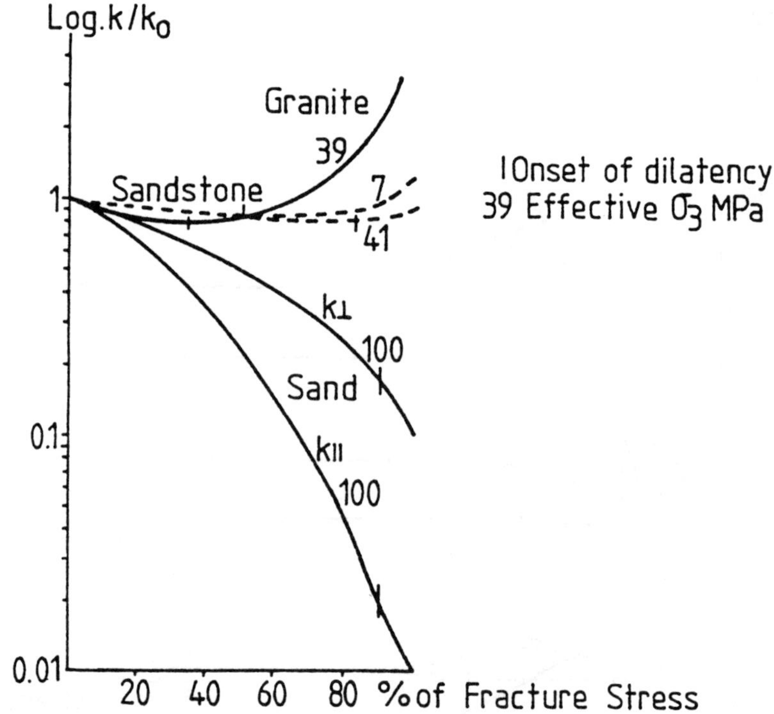

Figure 1.11 Permeability changes during testing to failure of Westerly Granite (Brace, 1978), Darley Dale Sandstone (Mordecai and Morris, 1970), and granite sand (Zoback and Byerlee, 1976). The hatches on the curves mark the onset of dilatancy; the numbers are the confining pressure in MPa.

fell monotonically. They also observed that the permeability of Ottawa Sand parallel to the major principal stress fell by a factor of 100 up to peak stress, while the permeability perpendicular to it was reduced by a factor of from 10 to 20%.

1.11 STRESSES AND DEFORMATION INDUCED BY EXCAVATION

When an underground structure is excavated in virgin rock, the equilibrium of forces in the rock mass, achieved over geologic time, is disrupted and the forces are redistributed. If the stresses redistributed around an opening exceed the bearing capacity of the rock mass, failure of the rock adjacent to the excavation boundary may lead to instability. Under normal circumstances this only occurs in the weakest rocks or for very large excavations. A more common rock deformation mechanism is by movement or loosening along existing discontinuities.

By applying the theory of elasticity to the behaviour of a rock mass, relations between induced and virgin stresses can be determined for various conditions of excavation geometry and internal support pressure. Although elastic analyses can only be completely satisfactory if the material behaves in an elastic manner over most of the stressed region, they clearly illustrate a number of important conclusions about conditions in actual rock masses.

The distribution of stresses around various shapes of opening is discussed in detail by Jaeger and Cook (1979), Hoek and Brown (1980) and Farmer (1985) among others. Various comments may be made, the most important being that in a hydrostatic stress field, around a typical rectangular opening, there will be stress concentrations at the corners and relatively high principal compression stresses parallel to the sidewalls. This is illustrated in Figure 1.12(a). The resultant effect is to cause shear fractures at the corners and tensile fractures parallel to the compression stresses (Figure 1.12(b)). In weak rocks these fractures appear as zones of increased permeability (Figure 1.12(c)) and the near field may be characterized as a series of dilatant tensile fractures parallel to the opening roof and sidewalls (Figure 1.12(b)).

It is evident that the resultant permeability of near-field strata around an excavation will be dependent on the interaction of a considerable number of processes and variables. Only approximate coupled analytical solutions to this problem are available. A certain amount of information concerning the general extent of the zone of permeability perturbations and the range of induced permeabilities is, however, available in the form of *in situ* test results, informed estimates and simple analytical approximations. In the vicinity of actual openings it is likely that permeability parameters will vary appreciably from any idealized 'average' both along and around the excavation.

Around a tunnel at a depth of 500 m in shale Heckman *et al.* (1979) suggest that a narrow but intensely fractured zone with an estimated hydraulic conductivity of 10^{-3} m/s and a porosity of up to 20%, i.e. approximately that of gravel, might exist. The average hydraulic conductivity of the general fracture zone was assessed at 10^{-6} m/s, with a range of from 10^{-5} to 10^{-8} m/s. Permeability changes in the floor and ribs were expected to be approximately one order of magnitude less than in the roof. Conditions around shafts are assumed to be somewhat better as gravitational failure is not a major problem, consequently fracture zones which develop are unlikely to be either as extensive or as highly disturbed. Values of hydraulic conductivity and porosity suggested by Heckman *et al.* (1979) are summarized in Table 1.1.

Permeability of strata within fracture zones was assumed to decrease exponentially with distance from a tunnel or shaft and effective permeabilities of fracture zones were therefore computed by multiplying the peak permeability at the excavation perimeter by a reduction factor, assumed in this instance to be 0.2. Values presented in Table 1.1 are the maximum Heckman *et al.* (1979) think would be reached in less than 30 to 100 years. They do, however, suggest that creep precipitation and recrystallization may, in time, reduce the permeability of fracture zones in salt to 1% of its maximum value. The permeability of fractured shale was expected to remain essentially constant with time.

The hydraulic conductivity of the rock immediately surrounding the Straight Creek tunnel pilot bore, at a depth of 450 m, was estimated from the measured reduction in pressure head through 2.4 m of rock and the flow rate along a length of the bore

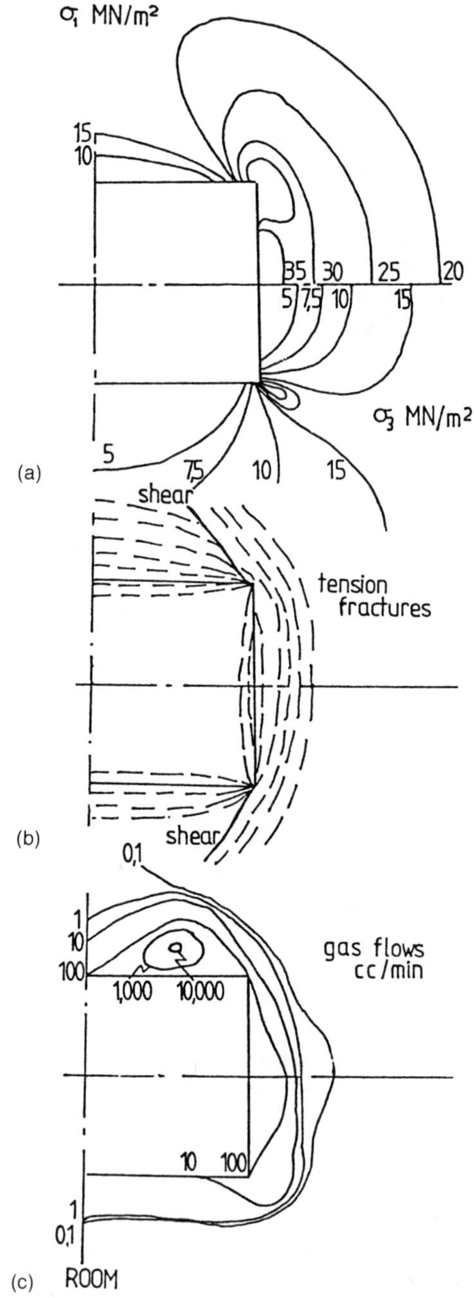

Figure 1.12 (a) Principal stress distributions around a hypothetical opening in an elastic material at a depth of 680 m; (b) rock fracture regime postulated around a similar opening by Gramberg and Roest (1984); (c) measured gas flows around a similar opening in rock salt at a depth of 680 m by Stormont (1990).

Table 1.1 Fracture porosity and hydraulic conductivity of fracture zones in shale and salt (after Heckman *et al.*, 1979)

	Hydraulic conductivity (m/s)		Porosity	
	Mean	Range	Mean	Range
Intensely fractured zone (shale)	10^{-3}	10^{-4}–10^{-2}	10^{-1}	10^{-2}–2×10^{-1}
General fracture zone (salt and shale)	10^{-6}	10^{-8}–10^{-5}	10^{-3}	10^{-4}–10^{-2}

by Hurr and Richards (1974). The calculated value of hydraulic conductivity, about 5×10^{-5} m/s, compares with an average value of 2×10^{-7} m/s for the rock mass obtained from slug tests.

Montazer *et al.* (1982) describe the results of a study undertaken to define the spatial distribution of permeability within a 5 m thick envelope around the CSM/ONWI room. The facility was excavated by smooth wall blasting techniques in a migmatite, biotite gneiss at a depth of 100 m below the ground surface. The room is 30 m long, 3 m high and 5 m wide. Forty-five boreholes were drilled in and around the room shortly after its excavation. Forty-two of these were drilled from inside the room in six radial sets; the remaining three holes were drilled horizontally, parallel to the excavation axis along the entire length of the room. Steady-state air and water injection techniques were employed in conjunction with pulse and pressure decay tests to measure permeabilities. Data from transient air injection tests were used to calculate the 'fracture index', which results from an approximate solution to the equations of gas flow through porous media. This is shown by Barron (1978) to be theoretically inversely proportional to permeability.

The mean of the logarithm of permeability and inverse of the fracture index for four groups of radial boreholes are plotted in Figure 1.13(a). It is evident that a zone of comparatively high permeability exists within the first 0.5 m of all these boreholes; 0.5 m was therefore taken to be the extent of the region of blast induced damage. A comparison of results from the four borehole groups provides evidence that the highest permeabilities are in fact induced in the roof strata. Interestingly, three of the zones containing fractures in the longitudinal boreholes show a diminishing trend in permeability towards the room (Figure 1.13(b)).

Four possible explanations were considered:

1. natural variations in fracture characteristics;
2. modifications of fracture traits due to blasting;
3. shear displacements along fractures near the opening;
4. closure of fractures due to an increase in normal stress.

A preliminary analysis of *in situ* stress measurements tended to favour the fourth possibility.

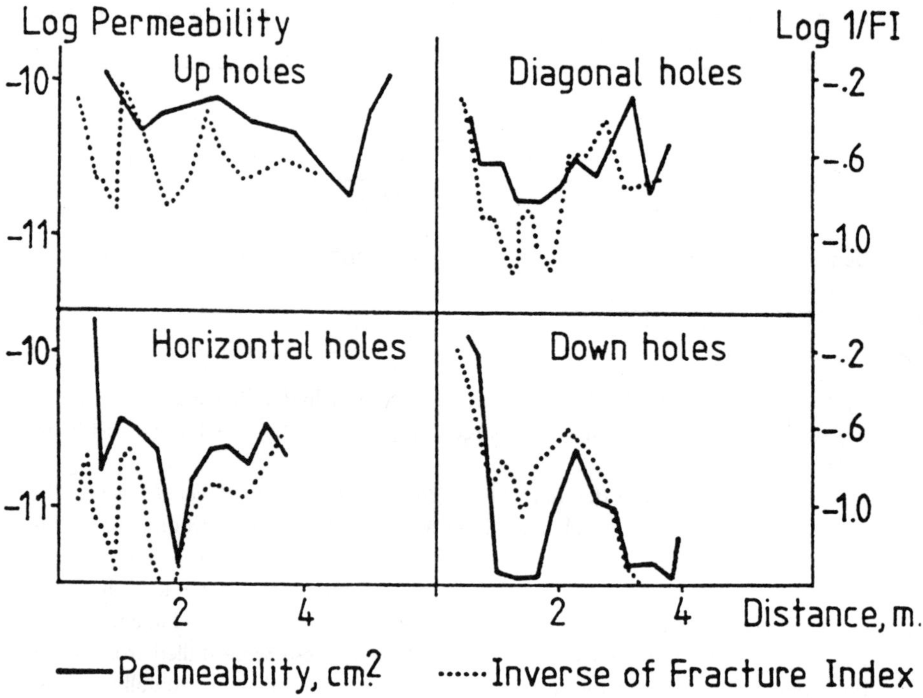

(a)

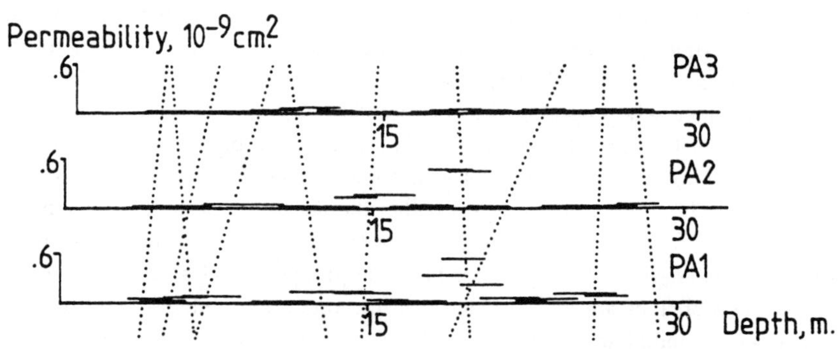

(b)

Figure 1.13 (a) Radial permeability and fracture index distribution in holes around a 3 m high by 5 m wide tunnel in gneiss and (b) permeability distribution in boreholes drilled parallel to a 3 m by 5 m × 30 m long opening in gneiss and 0.6 m from it (after Montazer *et al.*, 1982). *Note*: 1 darcy is approximately equal to 10^{-8} cm^2.

It was concluded that the zone of principal blast damage extends about 0.5 m from the excavation perimeter, while the regime of 'stress modification' extends further, possibly to 3.5 m. The permeability of fractures appears to be higher when they are orientated parallel to the axis of an excavation than when they are perpendicular to it.

1.12 PERMEABILITY CHANGES DURING MINING

Mine excavations represent an extreme case of high stress redistributions, often in weak rock, where significant imposed fracturing may occur. Whittaker and Singh (1978) studied permeability changes in a coal mine rib pillar behind an advancing longwall face by observing the pressure-flow characteristics of sections of experimental boreholes. Results obtained from a rib pillar, plotted as iso-conductivity lines, are illustrated in Figure 1.14.

The hydraulic conductivity of the pillar edge increased significantly as the distance from the face increased. Significant fracture dilation appears to have occurred at distances greater than 50 m from the face, final equilibrium being achieved at 120 m from the face. The results suggest that some phases of fracture dilation were followed by partial reclosure. Individual tests indicated a fair degree of uniformity. The effects of mining do not appear to have had a significant influence on the flow characteristics of this rib pillar at distances greater than 12 m from its edge. Measured hydraulic conductivities were generally found to be in the range 10^{-4}–10^{-6} m/s.

Fisekci and Barron (1975) analysed gas flow/pressure data obtained from in-seam boreholes bored ahead of longwall coal mine faces at sites in Alberta and Australia. Assuming Darcy's Law to be applicable and radial flow into borehole measuring stations, they applied the theory of unsteady radial flow of gases to express flow rate in terms of fracture permeability and effective fracture porosity. Figure 1.15 shows the variation of calculated permeabilities and porosities with distance into the

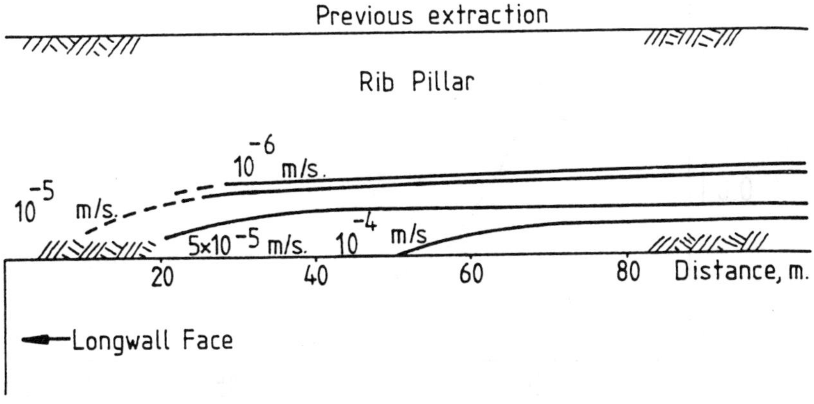

Figure 1.14 Distribution of hydraulic conductivity in a coal pillar adjacent to a longwall face (after Whittaker and Singh, 1978).

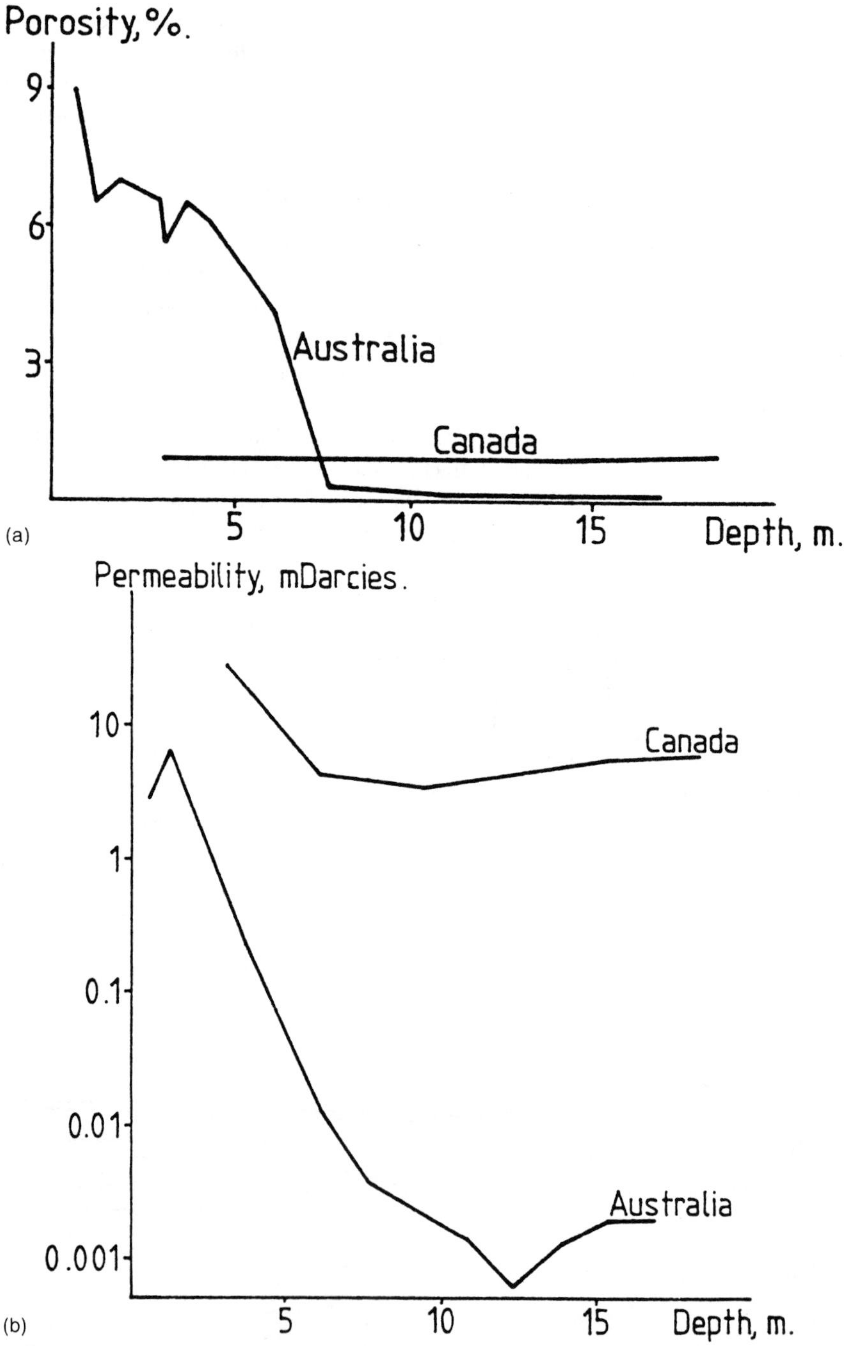

(a) (b)

Figure 1.15 Variation of (a) fracture porosity and (b) permeability with distance into a solid coal ribside behind an advancing longwall face (after Fisekci and Barron, 1975).

solid coal. The Australian longwall coal face data shows a variation in fracture porosity from approximately 9% close to the face to 0.2% about 7.5 m into the face, diminishing to a virgin coal fracture porosity of 0.1%. There is no indication of this type of variation at the Canadian site, although it may occur between 0 and 3 m from the solid edge where measurements were not taken. Virgin porosity at this site appears to be about 1%. Both sites showed relatively high fracture permeabilities close to the face, diminishing rapidly to a minimum 9–12 m from the face before rising again to approach the virgin permeability. The high permeabilities and porosities close to the excavation perimeters indicate the existence of highly fractured, permeable zones. The positions of minimum permeabilities were thought to occur where peak abutment stresses might be expected.

Air injection techniques have been used in a number of investigations into the relative fracture permeabilities of mine pillars and the intensity of fracturing in the strata around tunnels (e.g. Barron, 1978; Thomas, 1974; Miller *et al.*, 1974). Distinction between fractured and stress relieved strata (both high permeability) has not proved generally possible other than by a subjective judgement based on specific configurations. The validity of absolute permeabilities calculated from the results of such experiments is difficult to assess.

Swedish investigators acknowledge that a considerable number of factors may affect the permeability of the rock mass around a mine excavation. They also recognize the complexity of the interaction and the difficulties of providing a quantitative assessment of the resulting permeability. In order to estimate the effects of excavation on virgin rock permeability, they utilized an empirically derived relationship between vertical geostatic stress and horizontal strata permeability in conjunction with the stress distribution predicted by numerical analysis assuming 0° and 90° joint sets (Ratigan, 1977; Ratigan *et al.*, 1977; Lindblom *et al.*, 1977). Although simulation of the excavation did suggest that some regions around the opening would experience strength failure, no explicit account was taken of these regions in assessing final permeabilities.

Figure 1.16 illustrates the hydraulic conductivity characteristics of (a) roof, (b) sidewall and (c) floor strata resulting from tunnel construction based on the assumption that principal permeability directions and magnitudes are coincident with, and controlled by, the principal stresses. Although this may not in fact be the case, it is instructive in displaying theoretical permeability changes normal and tangential to the excavation. The calculated permeability normal, and close, to the tunnel periphery is low in comparison with the tangential component. Physically this can be explained in terms of an increased tangential compressive stress which reduces the permeability to a value below its initial magnitude. Normal stresses are zero along the tunnel perimeter but increase rapidly to undisturbed magnitudes. The result is high tangential permeabilities near the excavation which reduce to approximately pre-excavation levels about 3 m from the tunnel. This dimension would probably be greater if the regions experiencing failure were considered.

It is important to note that the permeabilities presented in Figure 1.16 are highly dependent on the assumed geostatic stress state and the excavation geometry. For example, if the rib were more nearly vertical, horizontal stresses in the floor would

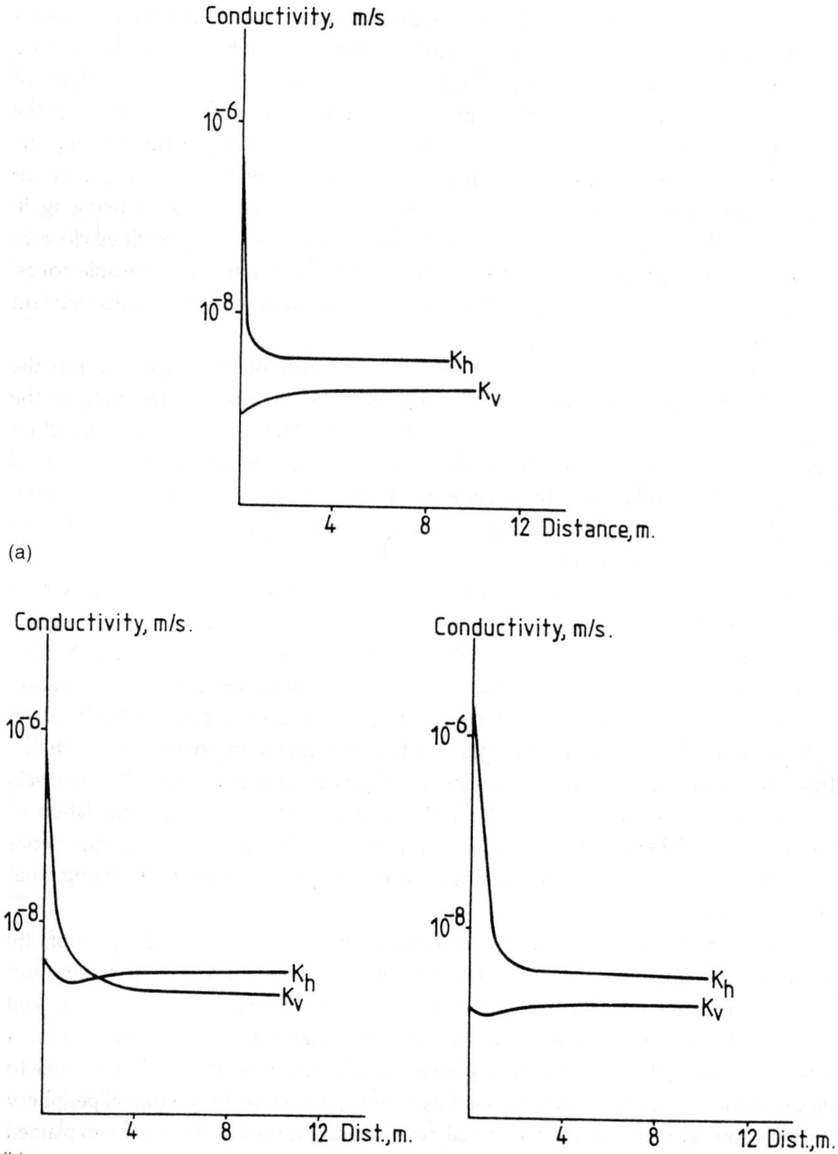

Figure 1.16 Vertical and horizontal conductivity in (a) the roof, (b) the sidewall, and (c) the floor of a circular tunnel based on the assumption that principal conductivity and stress directions are the same (after Ratigan, 1977).

be lower and vertical permeabilities greater. Similarly, if the ratio of horizontal to vertical stress had been taken to be unity rather than two, horizontal stresses in the roof would be lower and vertical permeabilities higher. A similar analysis assuming principal discontinuity sets at 45° and −45° yielded similar near-field permeability perturbations.

1.13 EFFECT OF EXCAVATION TECHNIQUE ON FRACTURE

Much more than by stress redistribution, rock disturbance during excavation is affected by the excavation technique. Disturbance can be very significant if the rock is blasted, less significant if excavation machinery is used.

When a quantity of high explosive is detonated in a borehole, high-pressure gases generated during the explosion impact upon the walls of the borehole, creating an intense pressure wave which propagates outwards into the rock. In the immediate vicinity of the borehole wall the induced stresses will exceed the strength of the rock by orders of magnitude, and crushing and shattering will occur. As the intensity of the explosion-induced stresses falls off rapidly with distance from the borehole due to geometrical dispersion, fragmented rock particle sizes will increase rapidly with distance from the borehole wall.

Beyond the zone in which the compressive strength of the rock is exceeded, a zone of primarily radial cracking will result from the tensile tangential stress component of the explosion-induced stress field assisted by venting of gases. Radial cracks continue to propagate as long as the tensile stress at the crack tip exceeds the tensile strength of the rock. In rock masses, radial fracturing is influenced by pre-existing discontinuities, the *in situ* state of stress and strata anisotropy.

A number of investigators have studied the extent of the zones of crushing and fracturing in the vicinity of blastholes. Comparison of their results is, however, complicated by the lack of a standard definition of these zones. In explosion technology, the effects of blasting are commonly scaled to the cube root of the explosive yield by the application of an empirically determined coefficient of proportionality. By specifying an explosive of a particular density, the radius of damage can then be directly related to the original charge (borehole) radius.

Cores taken in the vicinity of blastholes and examined in the laboratory using a variety of techniques have yielded considerable information on the effects of blast-induced micro-fracturing on mechanical and physical properties. Of six core damage analysis techniques employed by Siskind and Fumarti (1974) and Siskind *et al.* (1973), active acoustic sounding proved the most sensitive to fracture detection and offers promise for field application for differentiation between states of fracture. Strength measurements were not as definitive, the principal problem being scatter in the data. Porosity and permeability results provided problems in interpretation. Measured porosities showed both anomalously low values as well as high values within the zone of blast damage, possibly as a result of partial crushing or blocking of pores with blast-produced fines. Permeability measurements made on 2.5 mm long core pieces proved useful indicators of microscopic fracture density. Macroscopic

Table 1.2 Relative radii of crushed zone (R_c) and damaged zone (R_D) around blast holes of radius a

Rock	Explosive	$R_{c/a}$	$R_{D/a}$	Source
Granite	C-4	–	18–20	Olson *et al.* (1973)
Shale	60% Dynamite	–	42–55	Siskind *et al.* (1973)
Shale	AN-FO	–	15–22	Siskind *et al.* (1973)
Tuffaceous and pyroclastic	60% Dynamite	3	20–30	Cattermole and Hanson (1962)
Hard rock	–	–	20–23	Siskind and Fumarti (1974)
Soft rock	–	–	26–29	Siskind and Fumarti (1974)
Granite	TNT nuclear simulation	1.9	4.9	Derlich (1970)
Granite		3.0–4.5	–	Atchison and Tournay (1959)
Granite	C-4	2.3	–	D'Andrea *et al.* (1970)
Granite	AN-FO	<8	14	Siskind and Fumarti (1974)
Siltstone	Calculation	4.5	63	Drukovanyi (1976)

permeabilities would be considerably higher and valuable information could be gained from *in situ* permeability measurements in a zone of blast-damaged rock. Data from various sources is summarized in Table 1.2.

A number of investigators have studied the extent of the zone of (blast) damage around underground openings, the most widely used approach being by geophysical profiling of the strata in the vicinity of the excavation. Scott *et al.* (1968), for example, examined changes in the physical properties of rock surrounding the Straight Creek tunnel pilot bore. Employing seismic refraction and electrical resistivity techniques they found an altered zone extending up to about 5 m from the tunnel wall. The first metre or two of this zone was attributed principally to the effects of blasting, while deeper changes were considered to result from stress redistribution.

Manev and Avramova-Tacheva (1970) observed an increased joint intensity in the vicinity of a mine opening attributable, at least in part, to disturbance caused by the drill and blast method of construction extending to about 15 m from the opening. A similar increase in joint intensity was apparent in the vicinity of – and adjacent to – ore veins.

To compare the degree of strata disruption resulting from drill and blast and machine excavation techniques, information is required from sites where tunnels have been constructed in the same rock under similar stress conditions using both techniques. In the absence of detailed studies directly related to the degree of disturbance, comparisons may be inferred from observations of opening stability and fracture characteristics.

1.14 INFLUENCE OF THERMAL STRESSES ON PERMEABILITY

The influence of heat on rock behaviour, although important to geologic repository design, is a specialist area and is not considered in detail in the present text. A brief mention of the basic effects is, however, useful.

Increasing the temperature of a rock mass causes a volumetric expansion of the blocks of intact material which results in the constriction of discontinuities and an overall decrease in the effective rock mass permeability. Although it has been inferred, for example, that the permeability of the walls of underground petroleum storage caverns is lower when the oil is heated, there have been few quantitative studies of the phenomenon (Acres, 1978), and detailed data awaits further experimentation in the repository programs (Vieth *et al.*, 1985).

An interesting study was conducted in a granitic rock mass at a depth of 360 m in the Stripa Iron Ore Mine, Sweden (Lundstrom and Stille, 1978). At two sites the radial flow from 16 boreholes, around a 3 m radius circle, to a central hole were monitored as the rock mass temperature was increased by 25° C by circulating warmed water. The effect was to reduce the hydraulic conductivity of the rock from $4 \pm 0.8 \times 10^{-11}$ m/s to $1.8 \pm 0.3 \times 10^{-11}$ m/s. However, consideration of the change in viscosity of the permeant at the higher temperatures indicates that the intrinsic permeability of the rock mass had been reduced by a factor of approximately four. In a related study, water flow into peripheral boreholes was measured for the duration of a heater test (Carlsson, 1978). Water inflow reduced as a function of time, implying that conducting fractures were closing during the test, although monitored displacements of major fractures on the tunnel floor at the test site indicated dilation without exception, albeit of a small magnitude, to a maximum of 13.5×10^{-6} m. Three months after the heater was turned off, water inflow continued to decrease with time, indicating a time-dependent component of the *in situ* permeability response of a rock mass to heating. At the same time the apertures of a number of monitored fractures had still not closed to their pre-heating levels.

Barton and Lingle (1982) discuss the results of tests made *in situ* on an 8 m^3 block of jointed gneiss. The test joint exhibited a fourfold reduction in permeability when loaded from 0 to 6.9 MPa under ambient conditions and a thirtyfold reduction when temperature was also increased to 74° C. Increased temperature alone, with no change in the normal stress, reduced permeability tenfold.

One consequence of increasing the temperature within a salt formation is that brine-filled cavities tend to migrate up the temperature gradient towards the heat source. Brine in salt is found in quantities ranging from small crystal inclusions to very large pockets; typically the moisture content of rock salt ranges from 0.02% to more than 2%. Brine migration is attributed to the difference in the solubility of salt in brine between the warmest and coldest sides of the brine cavity. Under the influence of a temperature gradient, salt ions move from the warmest side of the cavity and are deposited at the coldest, the net effect being an apparent migration of the cavity up the temperature gradient. Bradshaw and McClain (1971) modelled the phenomenon as a pure diffusion process and concluded that the rate of cavity migration was an increasing function of temperature and directly proportional to the temperature gradient. Experimental data was found to agree with the diffusion model for temperatures up to 200° C.

Rock discontinuity characterization

2

It has been shown in Chapter 1 that the hydraulic behaviour of a rock mass is determined by the geometry of the fracture system, whether natural or imposed. Thus, in order to analyse the hydraulic behaviour of massive rocks, the first requirement is a detailed description of the characteristics and geometries of individual discontinuities and of the discontinuity system. This is the meeting point between conventions of rock mechanics and geohydrology. Rock mechanics – concerned with the stability of exposed surfaces and structures in massive rock – also requires a systematic description of discontinuities. Data on which this description is based is usually collected from one or two dimensional scanline surveys or cell mappings of borehole cores, exposures and tunnel walls, by individual engineers or geologists. Although it is an inexact method of data collection, reliability can be increased by statistical and probabilistic methods. Over the past ten years stochastic approaches based on distributions have been developed and these are an important step.

The mechanical properties of discontinuities are also important, because the aperture of a discontinuity through which it can transmit fluid, is determined by the surface roughness, the presence of infill, and the joint wall deformation and strength characteristics. The latter are particularly important, since the geostatic stresses at various depths of burial will critically affect the relative closure and openness of discontinuities.

2.1 DESCRIPTION OF DISCONTINUITIES

The International Society of Rock Mechanics *Suggested Methods for Quantitative Description of Discontinuities in Rock Masses* (Brown, 1981) recommends **discontinuity** rather than fracture as the generic term for breaks in rock masses. The term **fracture** is more commonly used in flow analysis, and in the current volume both will be used. The *Suggested Methods* propose other useful definitions which are worth quoting, as follows.

Joints are breaks of geological origin in the continuity of a body of rock along which there has been no visible displacement. A group of parallel joints is called a set and joint sets intersect to form a joint system. Joints can be open, filled or healed.

Joints frequently form parallel to bedding planes, foliation and cleavage and may be termed bedding joints, foliation joints and cleavage joints accordingly.

Faults are fractures or fracture zones along which there has been recognizable displacement, from a few centimetres to a few kilometres. The walls are often striated and polished (slickensided) as a result of the shear displacement. Frequently rock on both sides of a fault is shattered and altered or weathered, resulting in fillings such as breccia and gouge. Fault widths may vary from millimetres to hundred of metres.

Discontinuity is the general term for any mechanical discontinuity in a rock mass having zero or low tensile strength. It is the collective term for most types of joint, weak bedding planes, weak schistosity planes, weakness zones and faults.

Ten parameters are selected to describe the characteristics of discontinuities:

1. **Orientation:** the attitude of a discontinuity in space. It is defined by the dip direction (azimuth) and dip of the line of steepest declination in the plane of the discontinuity.
2. **Spacing:** the perpendicular distance between adjacent discontinuities. The term normally refers to the mean or modal spacing of a set of joints.
3. **Persistence:** the discontinuity trace length as observed in an exposure. It may give a crude measure of the areal extent or penetration length of a discontinuity. Termination in a solid rock or against other discontinuities reduces persistence.
4. **Roughness:** the inherent surface roughness and waviness relative to the mean plane of a discontinuity. Both roughness and waviness contribute to the shear strength. Large scale waviness may also alter the dip locally.
5. **Wall strength:** the equivalent compressive strength of the adjacent rock walls of a discontinuity. It may be lower rock block strength due to weathering or alteration of the wall. It is an important component of shear strength if the rock walls are in contact.
6. **Aperture:** the perpendicular distance between the adjacent rock walls of a discontinuity, in which the intervening space is air or water filled.
7. **Filling:** any material that separates the adjacent rock walls of a discontinuity. It is usually weaker than the parent rock. Typical filling materials are sand, silt, clay, breccia, gouge and mylonite, but also included are thin mineral coatings and healed discontinuities such as quartz and calcite veins.
8. **Seepage:** any water flow and free moisture visible in individual discontinuities or in the rock mass as a whole.
9. **Number of sets:** the number of joint sets comprising the intersecting joint system. The rock mass may be further divided by individual discontinuities.
10. **Block size:** the rock block dimensions resulting from the mutual orientation of intersecting joint sets and resulting from the spacing of the individual sets. Individual discontinuities may further influence the block size and shape.

The difficulty inherent in characterization is that few of these parameters can be determined with precision, and most can only be estimated either empirically or through analogy.

2.2 DISCONTINUITY ORIENTATION – STATISTICAL ANALYSIS

Methods of discontinuity orientation measurement, presentation and analysis have been described elsewhere (Priest, 1985; Goodman, 1976; Einstein and Baecher, 1983; Dershowitz and Einstein, 1988). The orientation of discontinuities can be measured from cores or from exposures using one or two dimensional scanlines. Discontinuity orientations are usually defined by dip direction and dip. They can be plotted on stereonets which may be equatorial conformal or equal area projections, or polar conformal or equal area projections. Measurement involves sampling – objective or subjective – and considerable inaccuracy. For this reason, the clusters or sets of discontinuity orientations representing preferred discontinuity directions in a rock mass, induced or imposed by deformation regimes, are inherently unreliable.

Statistical analysis of each set is needed in order to determine the distributions which can be fitted to the measured data and to estimate the dispersion around the mean, the central tendency and the theoretical distribution of goodness of fit. There are several approaches to orientation distribution. These include Arnold's hemispherical normal distribution, Fisher's hemispherical normal and bivariate distribution, Bingham's bivariate normal and bivariate log-normal distributions (Fisher, 1953; Bridges, 1975; MIT, 1979; Einstein and Baecher, 1983; Kulatilake, 1988). Some of these mathematical distributions and maximum likelihood estimators are shown in Table 2.1. Dershowitz and Einstein (1988) compared field geological data from a variety of sources and came to the general conclusion that none of the currently used distributions provided a statistically acceptable fit although the Fisher, bivariate Fisher and Bingham distributions provided an approximate equal number of best fits.

The Fisher distribution is a relatively simple method for estimating the mean direction of a collection of poles (Fisher, 1953). The mean direction is defined as the direction of the vector sum of each pole. The magnitude of the mean direction is called the 'resultant', R. The probability density distribution of fracture orientations is assumed proportional to $e^{\kappa \cos\psi}$ where ψ is the angular displacement from the mean and κ is the dispersion parameter about the mean direction. The resultant vector, $\mathbf{R}$, is referred to as the mean vector. If the magnitude of R is less than N measurements, the fracture direction becomes dispersed. κ is represented as an index of the dispersion of the orientation. When all N measurements are in the same direction, the resultant $|R|$ will be equal to the length N (Figure 2.1). When $N-|R|<2$, κ can be estimated in the form:

$$\kappa = \frac{N-1}{N-|R|} \tag{2.1}$$

and $\kappa > 0$ is a measure of dispersion. The Fisher distribution is also called the 'spherical normal' distribution. It is analogous to the normal distribution in a plane. Kulatilake (1985) used a chi-square goodness-of-fit test to develop a Fisher hemispherical distribution, treating it as a bivariate distribution. κ can be estimated from $\kappa = N/(N-|R|)$ for the cluster analysis and has values of $\kappa > 3$.

Table 2.1 Orientation distribution forms (after MIT, 1979)

Distribution	Mathematical form	Maximum likelihood estimators
Arnold	$f(\theta,\phi)=\dfrac{e^{k\cos\phi}\sin\phi}{e^{k}-e^{-k}}$	$k=\dfrac{\lvert R\rvert}{N}$ R denotes resultant vector N denotes no. of joints k is a measure of dispersion
Bingham	$f(\theta,\phi)=C_1\,e^{(k_1\sin^2\theta+k_2\cos^2\theta)\sin^2\phi}\sin\phi$	$\begin{vmatrix} k_1 & & \\ & k_2 & \\ & & 0 \end{vmatrix}=F\left(1/2;\,3/2;\begin{vmatrix} \mu_1 & & \\ & \mu_2 & \\ & & \mu_3 \end{vmatrix}\right)$ $\lvert\mu\rvert$ is a matrix including eigenvalues of covariance matrix of direction cosine C_1 is a normalization constant k_1 and k_2 are dispersion parameters F is a confluent hypergeometric function of matrix argument $\lvert\mu\rvert$ (Mardia, 1972, p. 235)
Bivariate normal	$f(\theta,\phi)=(2\pi\sigma_\theta\sigma_\phi)^{-1}\exp\left\{-1/2\left(\dfrac{\bar{\theta}^2}{\sigma_\theta^2}+\dfrac{\bar{\phi}^2}{\sigma_\phi^2}\right)\right\}$	$\bar{\theta},\bar{\phi}=$ sample mean azimuth, dip $\sigma_\theta,\sigma_\phi=$ sample standard deviations
Bivariate Fisher	$f(\theta,\phi)C_1e^{(k_1\sin^2\theta+k_2\cos^2\theta)\cos\phi}\sin\phi$	C_1 is a normalized constant k_1 and k_2 are dispersion parameters along major and minor principle axis.
Fisher	$f(\theta,\phi)=\dfrac{e^{k\cos\phi}\sin\phi}{1-e^{-2k}}$	$k=\dfrac{\lvert R\rvert}{N}$ $R=$ resultant vector $N=$ no. of joints

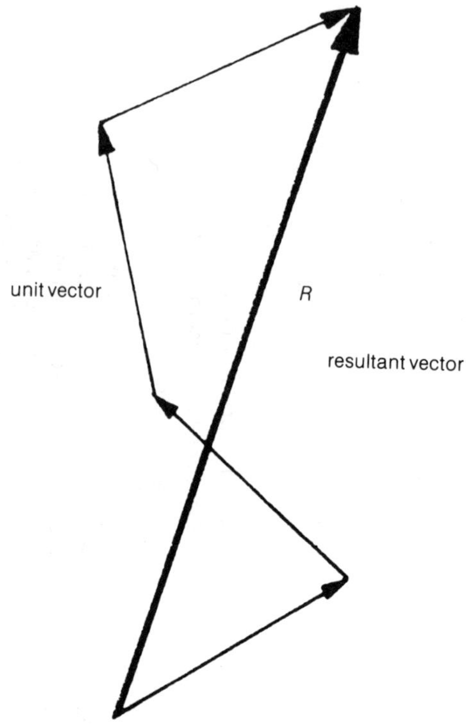

Figure 2.1 Fisher distribution method of determining the mean pole direction as the 'resultant' vector (**R**) sum of unit vectors representing individual poles.

To analyse clusters of orientation data, Mahtab *et al.* (1972) and Mahtab and Yegulalp (1984) developed a computerized method. They divided a sphere on which poles are plotted (Figure 2.2) into 100 equal-area quadrilateral patches. The density of the poles in each patch can then be obtained. They defined clusters as collections of all points in adjacent patches where each patch possesses a density which exceeds a threshold value. They then compared the density with a Poisson distribution to identify threshold values. When clusters have been identified, each cluster is compared with Arnold's hemispherical normal distribution with probability density $V(\psi, \kappa)$, given by:

$$V(\psi, \kappa) = \frac{\kappa}{4\pi(e^{\kappa} - 1)} e^{\kappa \cos\psi} \tag{2.2}$$

where ψ is the random variable which assumes that the value ψ_i is the angle between the ith observation and the mean vector and κ is a measure of dispersion. This distribution is similar to the univariate normal distribution. Mahtab *et al.* (1972) give estimates of κ for $\kappa > 6$ as $\kappa = N/(N - |R|)$. The probability p that a normal to a joint

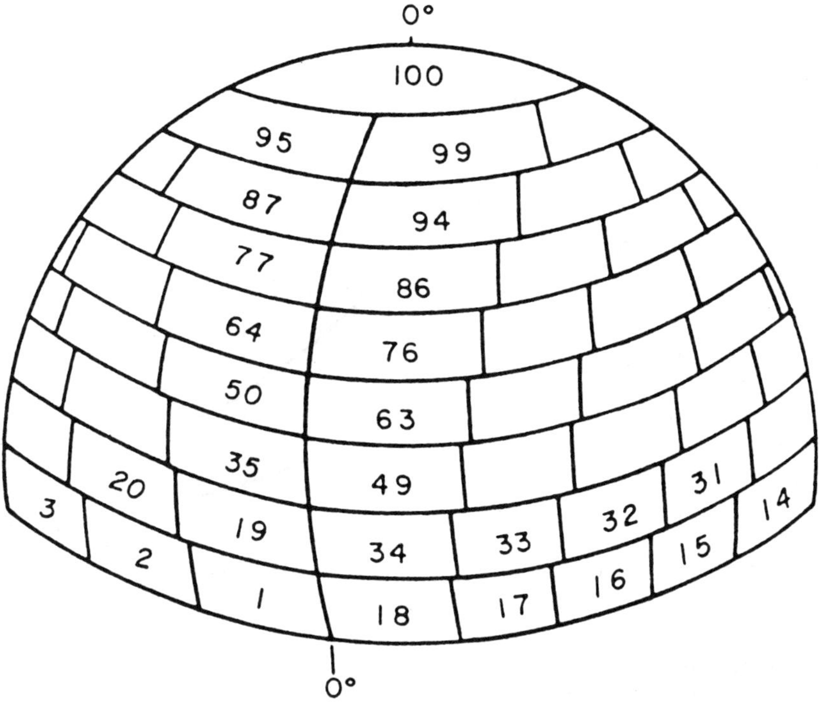

Figure 2.2 A method of analysing clusters of orientation data using an upper hemisphere surface divided into 100 equal area quadrilateral patches with a dip angle and azimuth range of nine bands (after Mahtab and Yegulalp, 1984).

Band no.	No. of patches	Dip angle range	Azimuth range
1	18	$90.00° \geqslant \phi \geqslant 79.63°$	20.0°
2	16	$79.63° \geqslant \phi \geqslant 70.12°$	22.5°
3	15	$70.12° \geqslant \phi \geqslant 60.66°$	24.0°
4	14	$60.66° \geqslant \phi \geqslant 50.95°$	25.7°
5	13	$50.95° \geqslant \phi \geqslant 40.54°$	27.7°
6	10	$40.54° \geqslant \phi \geqslant 30.68°$	36.0°
7	8	$30.68° \geqslant \phi \geqslant 10.95°$	45.0°
8	5	$19.98° \geqslant \phi \geqslant 8.11°$	72.0°
9	1	$8.11° \geqslant \phi \geqslant 0.00°$	360°

occurs within ψ of the mean orientation is given as:

$$\cos \psi = 1 + (1/\kappa) \ln(1 - p). \tag{2.3}$$

Angles of 50% and 95% probability poles from the mean of the hemispherical normal distribution are shown in Figure 2.3.

Since the orientation dispersion is an index of dispersive degree in space, the Fisher distribution can be used to calculate a mean vector from all the orientations

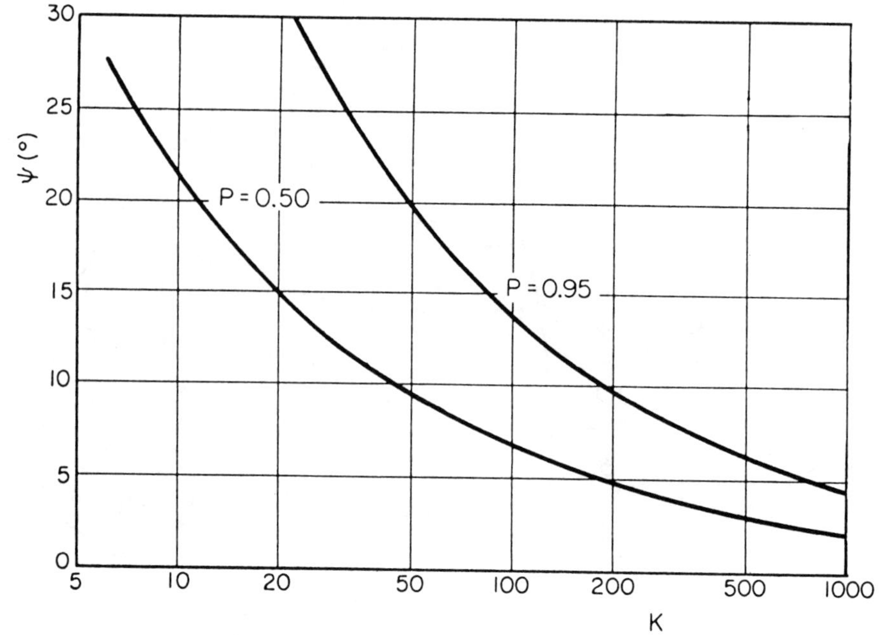

Figure 2.3 50% and 95% probabilities of a pole occurring within ψ of the mean pole direction (after Goodman, 1976).

in each set. The process of calculating the mean vector from a cluster or set has been described by Goodman (1976) and Kulatilake (1985). Each normal to a discontinuity is expressed as the direction cosine $(l, m, n,)$ and a unit vector. The resultant, **R**, of all the unit vectors in the set is the mean vector which is calculated by summing the direction cosine of each unit vector.

The spherical coordinates $(\theta, \phi, r = 1)$ can be related to dip angle ψ and dip direction α, given by:

$$\begin{aligned} \theta &= 2\pi - \alpha \qquad 0 \leqslant \theta \leqslant 2\pi \\ \phi &= \psi \qquad\qquad 0 \leqslant \phi \leqslant \tfrac{\pi}{2} \end{aligned} \tag{2.4}$$

$$\begin{aligned} l &= \sin\phi\cos\theta \\ m &= \sin\phi\sin\theta \\ n &= \cos\phi. \end{aligned} \tag{2.5}$$

If α_i and ψ_i are the dip direction and dip angle respectively of the ith observation in a set of N observations then the angular coordinates (θ_i, ϕ_i) of the normal to this point can be calculated from equation 2.4. The direction cosine (l_i, m_i, n_i) for this point can be calculated by equation 2.5. The resultant vector, **R**, of N of these observations passes through the mean vector of the clusters. The magnitude of the resultant vector,

$|R|$, is given by:

$$|\mathrm{R}| = \left| \left(\sum_{i=1}^{N} l_i \right)^2 + \left(\sum_{i=1}^{N} m_i \right)^2 + \left(\sum_{i=1}^{N} n_i \right)^2 \right|^{\frac{1}{2}} \tag{2.6}$$

and direction cosines for the mean vector $(\bar{l}, \bar{m}, \bar{n})$ are expressed as:

$$\bar{l} = \frac{1}{|R|} \sum_{i=1}^{N} l_i$$

$$\bar{m} = \frac{1}{|R|} \sum_{i=1}^{N} m_i \tag{2.7}$$

$$\bar{n} = \frac{1}{|R|} \sum_{i=1}^{N} n_i.$$

The alternative direction of the mean vector can be given by $(\bar{\theta}, \bar{\phi})$ as follows:

$$\bar{\phi} = \arccos(\bar{n})$$

or

$$\bar{\theta} = \arccos\left(\frac{\bar{l}}{\sin\bar{\phi}}\right) \qquad \text{if } \bar{m} \geqslant 0 \tag{2.8}$$

$$\bar{\theta} = 2\pi - \arccos\left(\frac{\bar{l}}{\sin\bar{\phi}}\right) \qquad \text{if } \bar{m} < 0.$$

Then the mean dip direction $= 360 - \bar{\theta}$ and $0 \leqslant$ the value of the arccos $\leqslant \pi$.

2.3 SPACING, FREQUENCY AND INTENSITY

The terminology for discontinuity spacing, frequency and intensity has been defined by many authors (Robertson, 1970; Hudson and Priest, 1979, 1983; Attewell and Farmer, 1976; Kazi and Sen, 1984, 1985). Discontinuity **spacing** is the distance between two parallel or sub-parallel fractures of a fracture set; **frequency** is the reciprocal of discontinuity spacing; **intensity** is the number of discontinuities per unit length, area or volume. **Rock Quality Designation** (RQD) is an empirical assessment of rock quality developed originally by Deere (1964) to describe the weathering of rock cores. It has subsequently been used extensively as a measure of discontinuity spacing. RQD is defined as the ratio of the total length (L_p) of the core fragments obtained over 100 mm (4 in) in length to the total length (L_t) cored, expressed as a percentage:

$$\mathrm{RQD} = 100 \frac{L_p}{L_t} \%. \tag{2.9}$$

Although RQD is often used in practical engineering reports, a disadvantage is that it does not take into account discontinuity aperture and condition. Figure 2.4

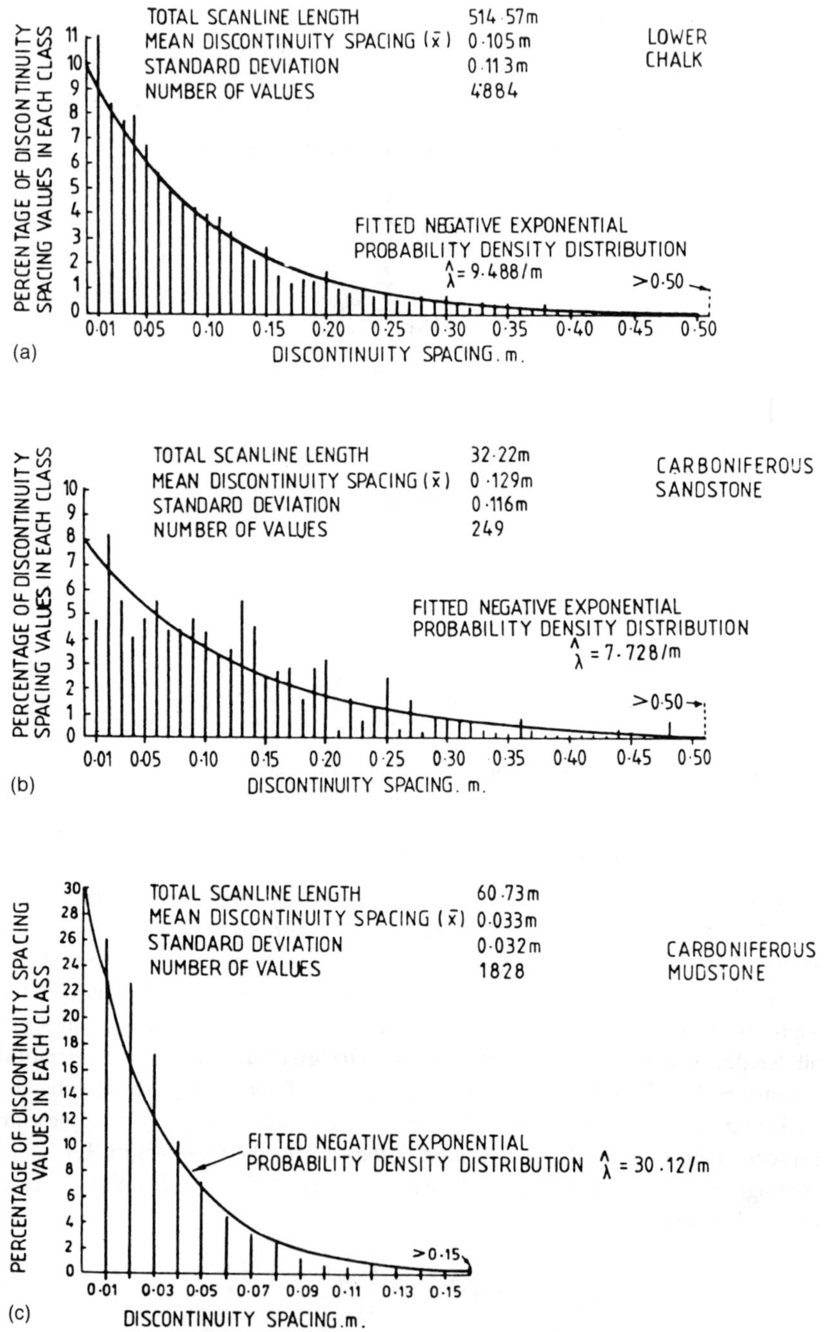

Figure 2.4 Typical discontinuity spacing histograms in chalk (a), sandstone (b) and mudstone (c) (after Priest and Hudson, 1976).

illustrates typical spacing distributions from the literature (Priest and Hudson, 1976). Although a wide variety of spacing distributions may be postulated, the majority fit a negative exponential or log-normal distribution. For instance, Call (1976), Priest and Hudson (1976), Baecher *et al.* (1977), Einstein *et al.* (1979), Wallis and King (1980) and Poulton *et al.* (1990), propose negative exponential distributions; Steffen (1975), Bridges (1975), Barton (1977), Baecher *et al.* (1977) and Sen and Kazi (1984) suggest log-normal distributions.

Priest and Hudson (1976) postulated that unless there is a large predominance of evenly-spaced discontinuities, any combination of clustered or random discontinuities will form a negative exponential distribution of discontinuity spacings. If discontinuity frequency is a negative exponential distribution, the relationship between RQD and frequency (number of discontinuities per unit length) can be expressed as:

$$RQD^* = 100e^{-\lambda l}(\lambda l + 1) \tag{2.10}$$

where RQD* is the theoretical RQD for the length criterion l, λ is the mean discontinuity frequency and l is an arbitrary threshold value (m). For the conventional RQD with a threshold of $l = 0.1$ m, the theoretical RQD can be expressed as:

$$RQD^* = 100e^{-0.1\lambda}(0.1\lambda + 1) \tag{2.11}$$

and if λ denotes the linear fracture frequency, RQD can also be expressed by a linear approximation:

$$RQD = -110.4 - 3.68\lambda \tag{2.12}$$

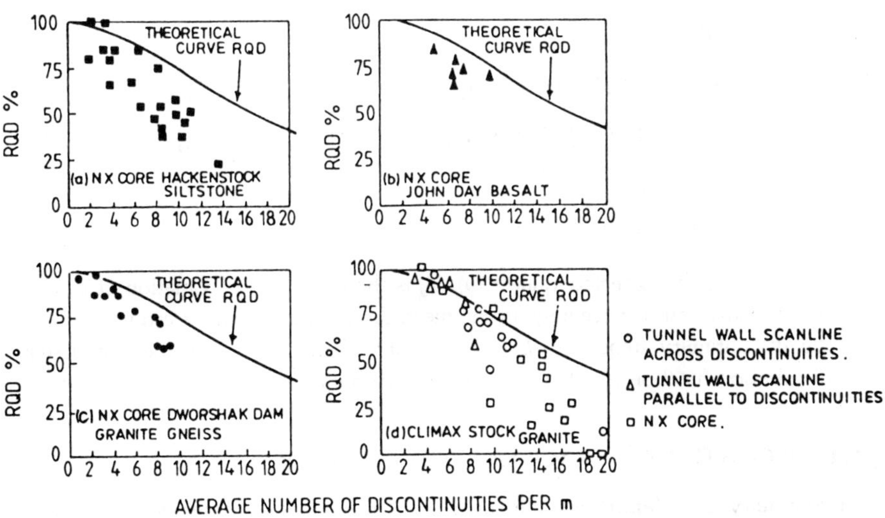

Figure 2.5 Relations between RQD and average discontinuity spacing for various sedimentary rocks with comparisons of actual data and theoretical curves (after Priest and Hudson, 1976).

Priest and Hudson (1976) show that at low RQD, the theoretical RQD tends to be higher than the measured RQD (Figure 2.5) because at low values the condition of the rock and the width of discontinuity openings will have a significant effect on measured RQD.

The theoretical RQD is also related to the orientation of the scanline. If the spacing and orientation of each discontinuity set are known, the discontinuity frequency along a scanline λ_θ, is related to the scanline orientation by:

$$\lambda_\theta = \sum_{i=1}^{n} \lambda_i \cos\theta_i \tag{2.13}$$

where λ_θ is the mean fracture frequency along a scanline, λ_i is the mean fracture frequency of the ith set along the normal to the ith set, θ_i is the acute angle between the scanline and the normal to the ith set, and n is the number of fracture sets.

Kazi and Sen (1984) derived an analytical expression for the prediction of discontinuity spacings and the relative error percent of mean discontinuity spacing estimation. They based this on a negative exponential and log-normal distribution of spacing. The expected values of spacing ($E'(y)$) and relative error percentage ($\alpha(y)$) can be calculated (Kulatilake, 1985, modified from Kazi and Sen, 1985) from:

1. negative exponential distribution:

$$E'(y) = \frac{1}{\lambda}\frac{1}{(1-e^{-\lambda L})} \tag{2.14}$$

$$\alpha(y) = 100\left\{1 - \frac{1}{1-e^{-\lambda L}}[1-(1+\lambda L)e^{-\lambda L}]\right\} \tag{2.15}$$

2. log-normal distribution

$$E'(y) = \bar{M}_y e^{\frac{1}{2}\xi_y^2}\{F[\ln(1/M_y)/\xi_y - \xi_y]/F[\ln(1/M_y)/\xi_y]\} \tag{2.16}$$

$$\alpha'(y) = 100\{1 - F[\ln(1/M_y)/\xi_y - \xi_y]/F[\ln(1/M_y)/\xi_y]\} \tag{2.17}$$

$$\bar{M}_y = M_y e^{-(1/2)\xi_y^2} \tag{2.18}$$

$$\xi_y = \ln(V_y^2 + 1) \tag{2.19}$$

where M_y is the expected value $E'(y)$, V_y is the coefficient of variation of y, $\alpha(y)$ is the relative error percentage in the mean discontinuity spacing estimation, λ is the discontinuity frequency, L is the scanline length and F is the cumulative distribution function for standard normal distribution.

2.4 DISCONTINUITY TRACE LENGTH AND SHAPE

Discontinuity **trace length** is a measurement of the two dimensional length of each discontinuity. In two dimensions, trace length is described as 'strike trace length' and 'dip trace length', and discontinuity size is referred to as 'joint area'. Since the exact interconnection of a rock mass is not easily determined the joint area can only be estimated. Discontinuity **shape** is a means of establishing the relation between the trace length and its orientation. Simplified discontinuity shapes are usually assumed – either circular, square, elliptic or rectangular (Robertson, 1970; Einstein and Baecher, 1983; Rasumssen *et al.*, 1985; Long and Witherspoon, 1985).

Robertson (1970) showed that the trace lengths of discontinuities can be fitted to exponential distributions when the shape is assumed circular. He concluded that discontinuity sizes are underestimated by trace lengths by a ratio:

$$A = \frac{16}{\pi^2} A' \tag{2.20}$$

where A' is the fracture area calculated from visible trace length and A is the average area.

The average fracture radius, r, is then

$$r = \sqrt{\left(\frac{A}{\pi}\right)} = \sqrt{\left(\frac{16A'}{\pi^2}\right)} = \frac{4}{\pi}\sqrt{\left(\frac{A'}{\pi}\right)} = \frac{4}{\pi}\frac{L}{2} \tag{2.21}$$

where $L/2$ is half of the expected magnitude of the trace length.

He also estimated the volume density of fractures, J_v, given the number of fractures, N, intersecting a simple line of length L as:

$$J_v = \frac{N}{LA\cos\delta\cos\theta} \tag{2.22}$$

where A is the average area of the joints and θ and δ are the horizontal and vertical angle between the sample line and the joint set pole. This estimate is based on the assumption that parallel joints are randomly distributed in space. The probability of intersecting N joints was assumed to follow a binomial law.

Since trace lengths are restricted to a sample in an exposure, the representative true trace length must be related to the observed trace length. When discontinuities intersect exposures such as tunnel walls, biases occur in mapping. These biases have been illustrated by many authors (Attewell and Farmer, 1976; Einstein and Baecher, 1983; Kulatilake, 1988). Particular causes of bias are:

1. The length of joint trace appearing in an exposure depends on the relative orientation between the exposure and the joint.
2. A larger discontinuity is more likely to appear in an exposure than a small one.
3. A larger discontinuity is more likely to appear than a shorter one in a sampling area on a scanline.
4. A truncation bias will result from an inability to recognize traces shorter than a certain threshold length.
5. A censoring bias will occur where it is impossible to measure the full lengths of traces that extend out of the field of vision. Possible measurement subsets are: both ends of the fracture, one end of the fracture and no ends of the fracture.

In order to increase the quality of statistical estimation and to predict the trace length distribution along any sampling area, as many as possible of the mapped traces should be used so that analytical formulae can be fitted to geometrical distributions of the sample region. If the sample region can be determined, the distribution of the mean trace length can be derived. Then the trace lengths of field data are first represented as a censored and truncated sample of this distribution. The predicted parameters of the true size distribution are obtained from this distribution through calculation of the statistical parameters of the complete trace distribution. As a result, although several forms of trace length distribution may be postulated, such as exponential, log-normal, hyperbolic or Gamma-1 distributions, data usually fits log-normal or exponential distributions, as shown in Figure 2.6. Studies by Robertson (1970), Call (1976), Cruden (1977) and Priest and Hudson (1981) support a negative exponential distribution; studies by McMahon (1974), Bridges (1975), Barton (1977), Baecher *et al.* (1977) and Einstein *et al.* (1979) support a log-normal. Based on an analysis of probability associated with given fractures, Pahl (1981) has investigated trace lengths in a specified mapping window. Pahl classified joints in the window into three types, depending on whether they intersect, transect or contain the joint trace. The mean trace length within non-vertical trace sets can be calculated by:

1. Log-normal distribution of true trace length

$$\mu = \frac{l(h - 2\eta)(1 + \theta_T - \theta_c)}{(l\cos\phi + h\sin\phi)(1 - \theta_T + \theta_c)} \tag{2.23}$$

2. Exponential distribution of trace length

$$\mu = \frac{l(h - \eta)(1 + \theta_T - \theta_c)}{(l\cos\theta + h\sin\theta)(1 - \theta_T + \theta_c)} \tag{2.24}$$

where μ is the population mean trace length, l is the length of mapping zone, h is the height of mapping zone, η is the lower cut-off length in the mapping zone, ϕ is the complement of the apparent dip (ϕ_a) of the joint set in the mapping zone where $\phi = 90° - \phi_a$, $\theta_T = N_T/N$ is the proportion of observed trace length that transects the mapping zone, $\theta_c = N_c/N$ is the proportion of observed trace lengths that are contained within the mapping zone, N_T is the number of traces transecting the mapping zone, N_c is the number of traces contained in the mapping zone and N is the number of traces intersecting the mapping zone.

By using the method of moments, Laslett (1982) derived the following distribution-free estimator for the mean trace length if the sampling zone is chosen as a circle in the planar cut:

$$L_1 = \frac{2D}{(2n + 2m)}$$

$$L_2 = \frac{D}{N} \tag{2.25}$$

$$L_3 = \frac{\pi SD}{[(n + m + p)\pi S - 2D]}$$

$$D = \sum_{i=1}^{n} X_i \sum_{i=1}^{m} Y_j + \sum_{k=1}^{p} Z_k$$

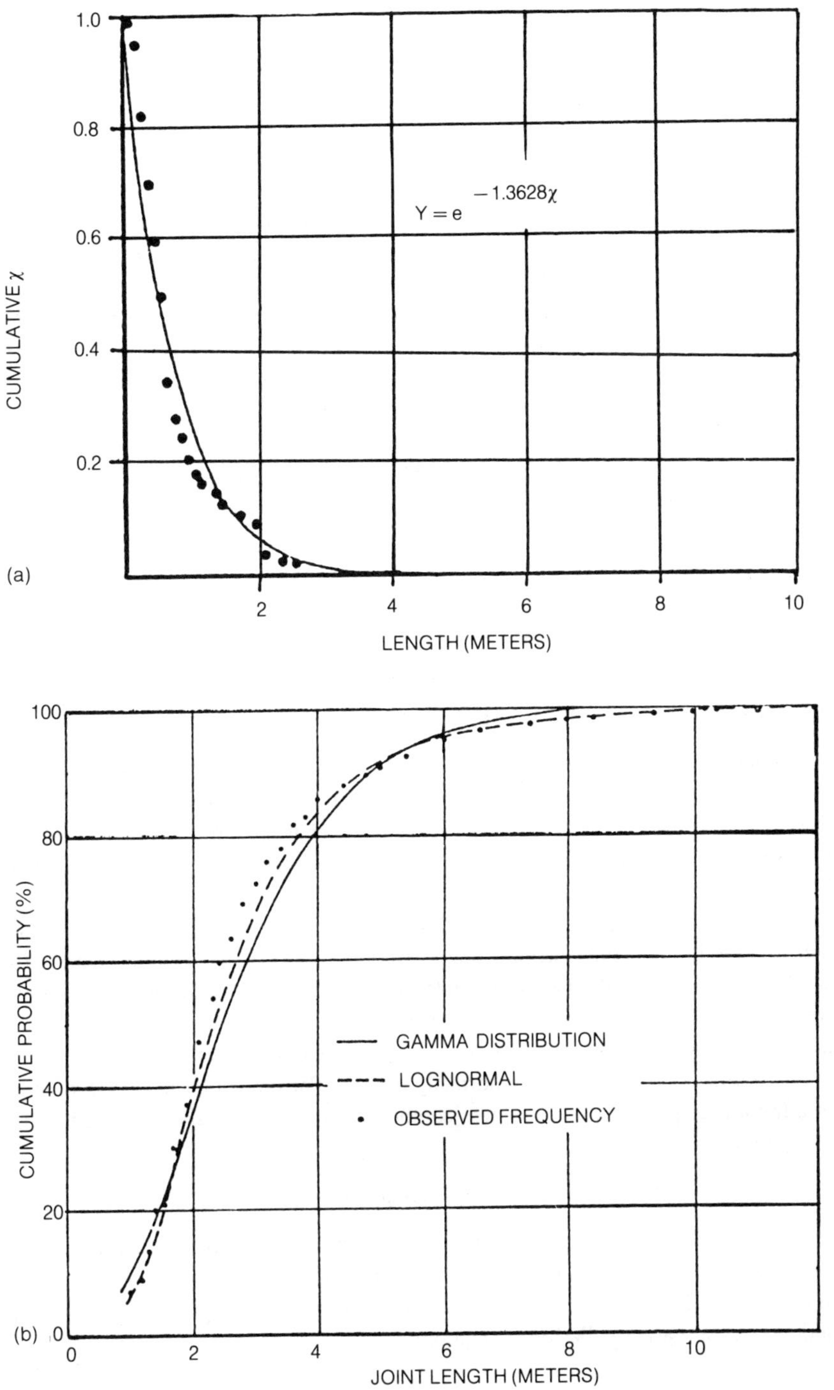

Figure 2.6 Discontinuity trace length histograms fitting (a) exponential distribution and (b) lognormal or gamma distribution (after Einstein *et al.*, 1979).

where n is the number of uncensored traces (both ends are observed), m is the number of semi-censored traces, p is the number of censored traces, X is the length of an uncensored trace (both ends observable), Y is the length of a semi-censored trace (one end observable), Z is the length of a censored trace (no end observable), S is the radius of the sampling and N is the number of observed points, defined as points on the segment (i.e. its southern end point). For a larger data set, the three estimators L_1, L_2 and L_3 should give the same results. In fact, the results will be different, and this eventual difference can be adjusted as a quality measure of the estimators.

In practical engineering, a borehole is the preferred method for obtaining geological data. Since borehole discontinuity logs can only correlate discontinuity intersections in closely spaced boreholes, a large number of boreholes are required to delineate the areal extent of discontinuities. The cost will limit the determination of the areal extent of fractures. On the other hand, trace length is often used to represent the areal extent when it is observed in a surface trace. It cannot, however, represent the true discontinuity shape or size. Thus, the analysis of discontinuity areal extent from discontinuity trace measurements is possible only if certain simplifications can be made. For instance, the distribution of discontinuity centroids can be assumed random; discontinuities can be assumed to be circular, elliptical or polygonal and distribution of discontinuity lengths can be assumed to be exponential or log-normal.

2.5 DISCONTINUITY ROUGHNESS

Roughness is usually considered solely in relation to the angle of sliding friction, ϕ, along a discontinuity in conventional rock mechanics. In considering flow through discontinuities, roughness is important for the effect it has on the aperture. As will be seen later, the conventional descriptions of roughness are a useful tool in fluid flow analysis.

The concept of roughness is illustrated in Figure 2.7(a) (after Barton, 1973) which defines descriptive terms such as rough (irregular) stepped (I), smooth stepped (II), slickensided stepped (III), rough or irregular undulating, smooth undulating and slickensided undulating and rough, smooth and slickensided planar. These are the bases for Barton's (1973) empirical description of joint roughness profiles.

Discontinuity shear strength is normally defined (Patton and Deere, 1970) in terms of a friction angle ϕ and a roughness component (i). Then the shear strength τ of a discontinuity is given by:

$$\tau = \sigma' \tan(\phi + i) \tag{2.26}$$

where σ' is the effective normal stress. ϕ may be either peak or residual, but for most stability purposes the residual or basic friction angle value ϕ_r or ϕ_b is included in equation (2.26). This has a value for most rocks of about 30°. Then $\phi_{peak} = \phi_r + i$ and this is the basis for Barton's equation:

$$\phi_{peak} = \text{JRC} \log_{10}\left(\frac{\text{JCS}}{\sigma'}\right) + \phi_r \tag{2.27}$$

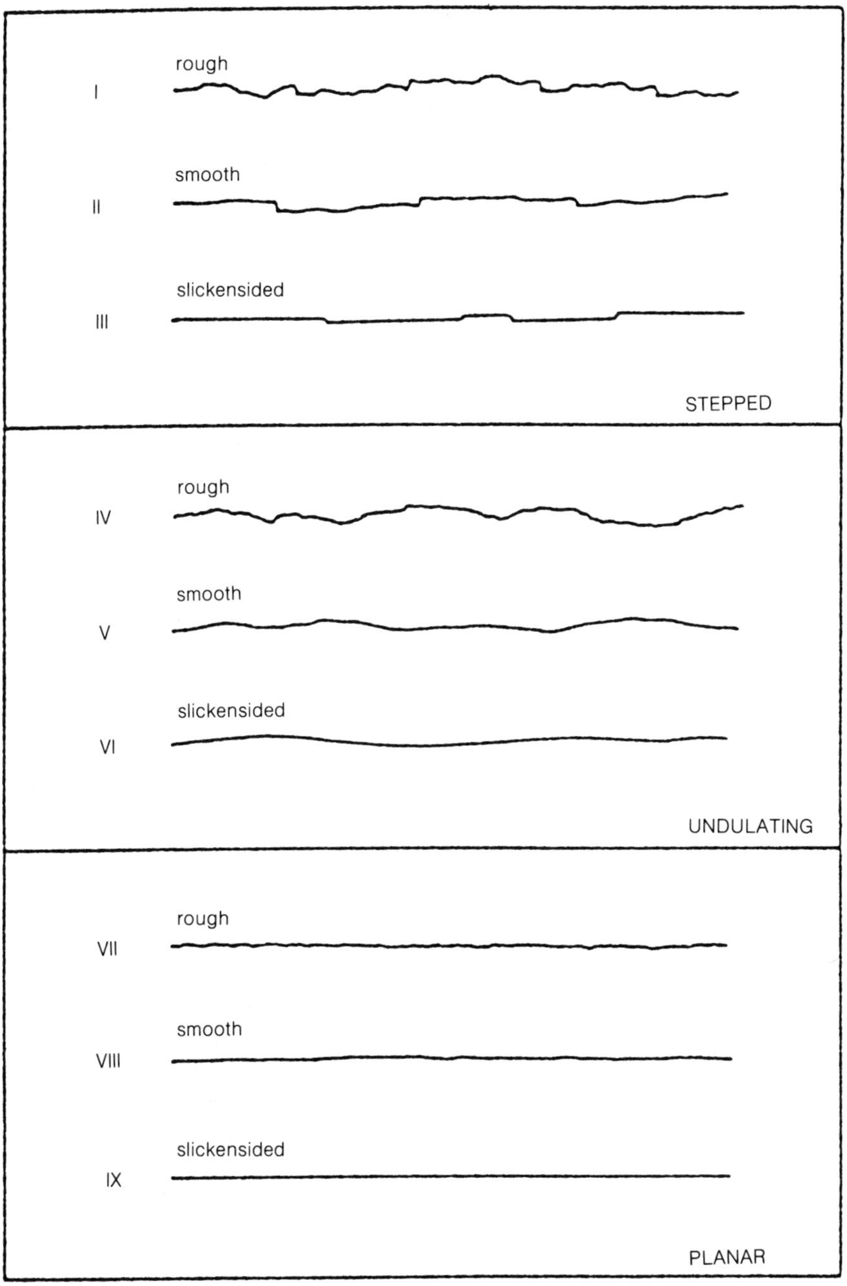

(a)

Figure 2.7 Typical roughness profiles: (a) classified according to surface properties (lengths 1–10m) and (b) with recommended JRC range (length 0.1m). After Barton, 1973.

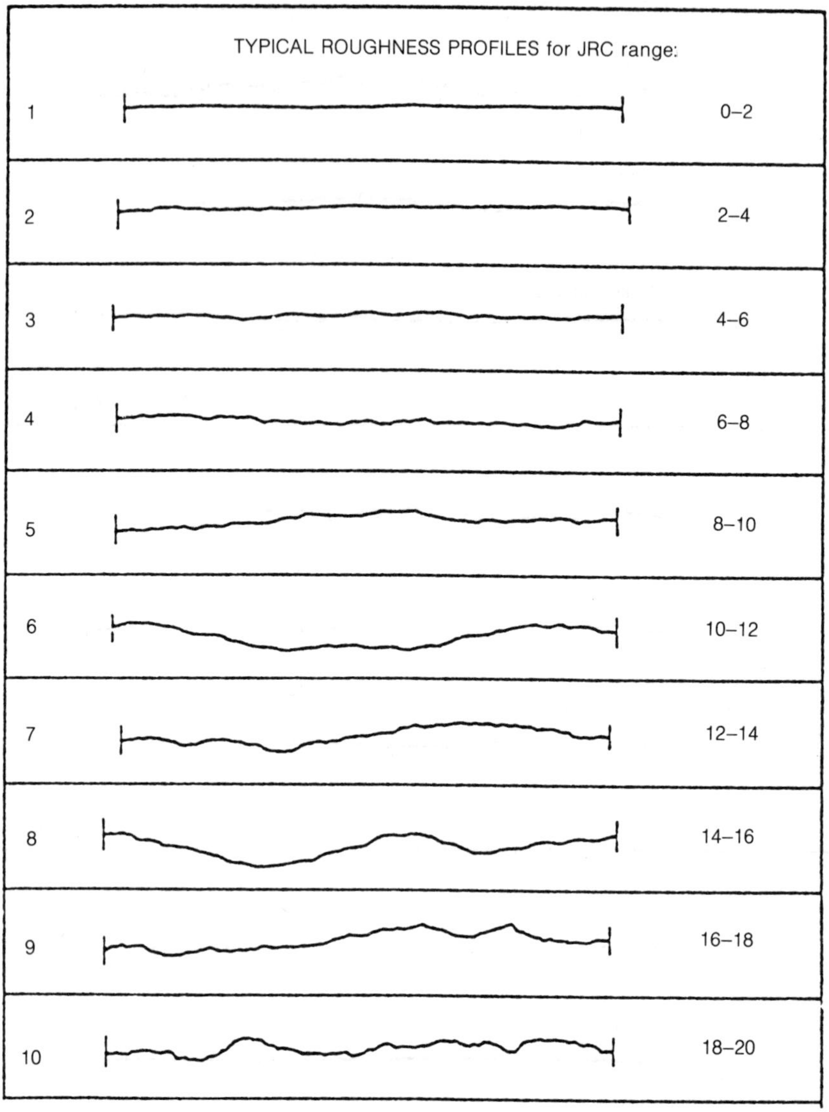

(b)

Figure 2.7 *contd.*

where JRC is the Joint Roughness Coefficient illustrated in Figure 2.7(b) and JCS is the Joint Wall Compression Strength.

The application of the method is illustrated in Figure 2.8. The roughness profile which most fits the profiles in Figure 2.7(b) is used to estimate JRC. Then the joint walls are tested with a Schmidt hammer to obtain a value of JCS. ϕ_r is assumed

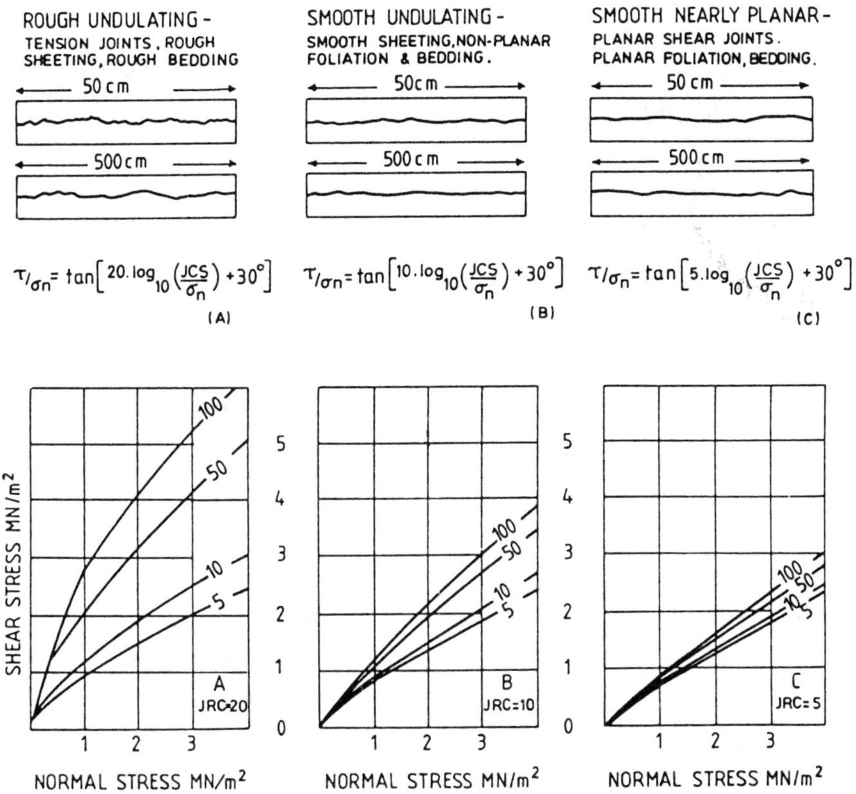

Figure 2.8 Illustrations of peak shear strength envelopes based on Barton *et al.*'s (1974) interpretation of JRC and JCS.

equal to 30°. It is a surprisingly accurate method of estimating ϕ_{peak} or i and as will be seen later JRC and JCS are very useful in estimating aperture closure.

2.6 WALL STRENGTH

Barton (1973) suggested several methods for estimating wall strength. The most accurate is to carry out a Schmidt hammer test. The device, which measures rebound hardness, and its use, is described incorrectly in Brown (1981), more accurately but in a special application by Poole and Farmer (1980). The value of rebound hardness can then be converted to compressive strength using Figure 2.9. Provided the highest of ten impacts at a single point is taken with the hammer normal to the surface (ideally vertically downwards) the results are reliable.

A less satisfactory, but very useful, method is to use the ISRM chart which gives a sensible classification of rock strengths. This is included in Table 2.2.

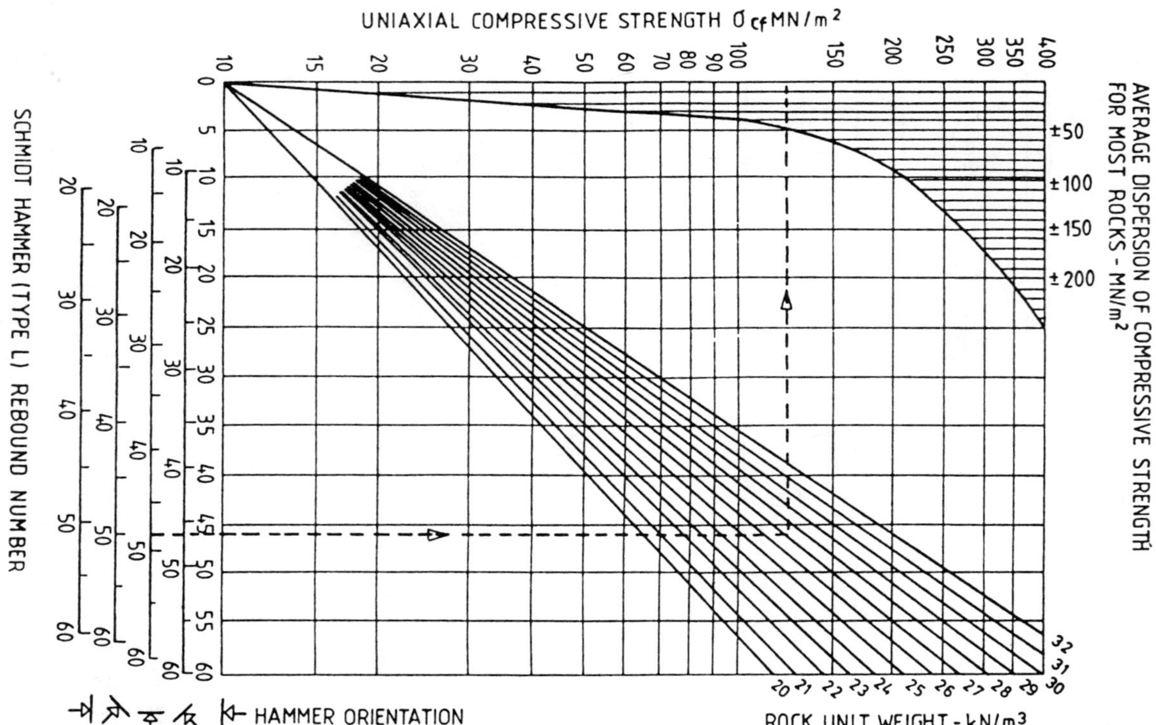

Figure 2.9 Schmidt hammer correlation chart relating rock unit weight, hammer rebound number and compressive strength (after Deere and Miller, 1966).

Table 2.2 Classification of joint wall compression strength (after Barton, 1973)

Grade	Description	Field identification	Approx. range of uniaxial compressive strength (MN/m^2)
S1	Very soft clay	Easily penetrated several inches by fist	<0.025
S2	Soft clay	Easily penetrated several inches by thumb	0.025–0.05
S3	Firm clay	Can be penetrated several inches by thumb with moderate effort	0.05–0.10
S4	Stiff clay	Readily indented by thumb but penetrated only with great effort	0.10–0.25
S5	Very stiff clay	Readily indented by thumbnail	0.25–0.50
S6	Hard clay	Indented with difficulty by thumbnail	>0.50
R0	Extremely weak rock	Indented by thumbnail	0.25–1.0
R1	Very weak rock	Crumbles under firm blows with point of geological hammer, can be peeled by a pocket knife	1.0–5.0
R2	Weak rock	Can be peeled by a pocket knife with difficulty, shallow indentations made by firm blow with point of geological hammer	5.0–25
R3	Medium strong rock	Cannot be scraped or peeled with a pocket knife, specimen can be fractured with single firm blow of geological hammer	25–50
R4	Strong rock	Specimen requires more than one blow of geological hammer to fracture it	50–100
R5	Very strong rock	Specimen requires many blows of geological hammer to fracture it	100–250
R6	Extremely strong rock	Specimen can only be chipped with geological hammer	>250

Note: Grades S1 to S6 apply to cohesive soils, e.g. clays, silty clays, and combinations of silts and clays with sand, generally slow draining. Discontinuity wall strength will generally be characterized by grades R0–R6 (rock) while S1–S6 (clay) will generally apply to filled discontinuities

Some rounding ot strength values has been made when converting to SI units.

2.7 DISCONTINUITY APERTURE

Apertures have importance mechanically, but are the critical factor in determining flow. In rock mechanics, they are usually classified by size as in Table 2.3.

Table 2.3 Aperture classification (after Barton, 1973)

Aperture (mm)	Class
<0.1	Very tight
0.10–0.25	Tight
0.25–0.50	Partly open
0.50–2.50	Open
2.50–10.0	Moderately wide
<10	Wide

Apertures are difficult to define in terms of true width, since the asperities which create joint roughness also affect the aperture opening size. Contacts between these asperities also affect the average opening size. For this reason, when considering transmissivity, apertures are normally defined in terms of their flow properties as equivalent parallel plate apertures subject to viscous incompressible flow (Snow, 1968a). Then if the flow is laminar and the plates horizontal, the flow velocity–hydraulic gradient relation becomes:

$$v=\left(\frac{\gamma}{12\mu}\right)a^2i \tag{2.28}$$

where V is the flow velocity, γ is the unit weight of the fluid, μ is the dynamic viscosity of the fluid, a is the parallel plate aperture and i is the hydraulic gradient.

The volumetric flow rate per unit plate width will then be:

$$Q=\left(\frac{\gamma}{12\mu}\right)a^3i. \tag{2.29}$$

By direct analogy with Darcy's Law (equation 1.4) where $v=Ki$ and $Q=KA(\Delta h/L)$, the volumetric flow rate becomes:

$$Q=Ca^3\Delta h \tag{2.30}$$

where Q is the volumetric flow rate and Δh is the change in hydraulic head. This equation is generally known as the 'Cubic Law', with the constant C representing the fluid properties and the flow geometry of the system. For the case of straight flow:

$$C=\frac{W}{L}\left(\frac{\gamma}{12\mu}\right) \tag{2.31}$$

and for the case of radial flow:

$$C = \left(\frac{2\pi\gamma}{12\mu}\right)\ln\left(\frac{R}{r}\right) \tag{2.32}$$

where W is the plate width, L is the plate length, R is the outer radius of flow and r is the inner radius of flow.

The dimensionless Reynolds number (R_e) is often used to study the type of flow regime encountered in a discontinuity. It is a measure of the ratio of inertial forces to viscous forces of a flow system and involves velocity, fluid properties and a characteristic length:

$$R_e = v\frac{D}{\nu} \tag{2.33}$$

where V is flow velocity, $D = 4R_H$ is four times the hydraulic radius and ν is the dynamic viscosity. For flow between parallel plates the hydraulic radius is equal to one half of the aperture, and the Reynolds number becomes:

$$R_e = \frac{2va}{\nu}. \tag{2.34}$$

A number of workers have studied experimentally the validity of the cubic law in the laboratory. Louis (1969) investigated the effects of surface roughness on the flow regime encountered in the open fractures, and postulated that flow regime was a function of the Reynolds number for the flow, and a surface roughness index:

$$S_R = \frac{H}{d} \tag{2.35}$$

where H is the mean height of the asperities and d is the hydraulic diameter and equal to twice the fracture aperture.

In order to account for surface roughness and turbulence, similar empirical models have been developed or studied by Sharp and Maini (1972), Gale (1985) and Iwai (1976). Romm (1966) demonstrated the validity of the cubic law for his models down to apertures of 0.2 μm. The character of the flow paths at such apertures has long been debated and the question of whether the cubic law holds for fractures partially closed by contact points under confining stress was studied by Iwai (1976). In a paper by Witherspoon *et al.* (1980), the authors concluded that the cubic law was valid for apertures down to 4 μm, despite local areas of contact at stresses up to 20 MPa. They introduced a constant f, ($f \geqslant 1$) into the cubic law in order to account for deviations:

$$Q = Ca^3\Delta h\left(\frac{1}{f}\right). \tag{2.36}$$

The cubic law is considered further in the next chapter.

2.8 DISCONTINUITY INFILL AND SEEPAGE

These have minimal significance hydraulically and are included for completeness. Infill is the material separating the adjacent rock walls of the discontinuities – such as calcite, chlorite, breccia, fault gouge, clays and silts. Infill can replace part of the aperture with materials ranging in strength from the strongest to the weakest. The grades in Table 2.2 apply and the effects on aperture are the same as if there were a smaller aperture with the same wall material.

Seepage is effectively the subject of the book and results from the discontinuity geometry and its relation to groundwater conditions. The rate of seepage is roughly proportional to the local hydraulic gradient and to the relevant directional permeability. In certain sedimentary rocks such as sandstones, the permeability of the rock between discontinuities may be significant so that a proportion of the total seepage occurs through the pores. In low permeability crystalline rocks such as granite, seepage will be through the discontinuities. In engineering, the seepage pressure or force induced in pores or fissures by flow may have mechanical significance and is given by:

$$F = \gamma_w i A \Delta L = \gamma_w i V \tag{2.37}$$

where i is the hydraulic gradient, A the discharge area, ΔL the flow path length and V the volume affected.

2.9 JOINT SETS AND BLOCK SIZE

The number of individual joint sets and the blocks bounded by them are of critical importance in any consideration of flow. Fundamental to any understanding of the role of joint sets is the concept of Representative Elementary Volume (REV). In order to obtain a valid or representative sample of a rock mass, the dimension of the sample must be sufficiently large to have properties – both hydraulic and mechanical – which are similar to those of the mass. In the case of rock strength, strength is related inversely to the square root of size (Farmer, 1983) and a sample size of several cubic metres is required before a REV is approached. In the case of flow, the size must be sufficient to include all joint sets and discontinuities. As discussed later in Chapter 5, in some rocks REV may not exist. The size of rock blocks and the number of joint sets gives an indication of the validity of REV.

Volumetric joint count (Palmstrom, 1982) is determined from the mean spacing of the joint sets as:

$$J_v = \frac{1}{s_1} + \frac{1}{s_2} + \frac{1}{s_3} + \cdots + \frac{1}{s_i} \tag{2.38}$$

where J_v is the volumetric joint count (number/cubic metre) and s_i is the mean spacing of the ith joint set in metres.

Kazi and Sen (1985) proposed that the relation between the volumetric fracture frequency, J_v (number of fractures per cubic metre) and the maximum number of matrix blocks, N_{bmax}, can be expressed as:

$$N_{\mathrm{bmax}} = \left(\frac{J_v}{3} + 1\right)^3. \tag{2.39}$$

They also proposed a volumetric RQD to describe the quality of rock masses. The relation between the volumetric RQD and the frequency of volumetric blocks N_b, is expressed by:

$$V.\mathrm{RQD} = 100e^{-0.001N_b}(0.001N_b + 1). \tag{2.40}$$

To estimate the relation between volume density and RQD, the International Society for Rock Mechanics (Brown, 1981) proposed that the relation between $|J_v|$ and RQD be expressed as:

$$\mathrm{RQD} = 115 - 3.3\,J_v \quad \text{and} \quad \mathrm{RQD} = 100\% \text{ if } J_v < 4.5. \tag{2.41}$$

A less satisfactory indicator is the **block size index**, I_b, which represents the average dimensions of typical rock blocks. The index can be estimated by selecting several block sizes and taking their average dimensions. Therefore the range of size domains in the rock mass will be represented by a series of block size indices.

A more satisfactory approach has been suggested by Mojtabai *et al.* (1989), Poulton *et al.* (1990) and Ghosh (1990), who use a scale invariance technique to define discontinuity structures and fragmented rock. Scale invariance requires a power law relation between a number and size. Such a relation is by definition a fractal. There are numerous examples of scale invariant processes in nature which can be quantified by the fractal concept (Mandelbrot, 1982). An example which has direct application to rock mechanics has been illustrated by Turcotte (1986). He showed that fragmented objects, including rock, coal, soil and meteoric fragments which had been broken in various ways, were all characterized by a size–frequency relation of the form:

$$N(r) \propto r^{-D} \tag{2.42}$$

where $N(r)$ is the number of fragments with a characteristic linear dimension greater than r and the relation is a fractal defined by the fractal dimension D. Fractal dimensions for fragmented geological materials are summarized by Turcotte (1986) and range from 1.14 to 3.54.

It has been conventional in the past to describe fragmention from blast products by a negative exponential dependence:

$$\frac{N(r)}{N} = \exp\left[-\left(\frac{r}{r_0}\right)^n\right] \tag{2.43}$$

where N is the total number of the fragments and r_0 is the mean fragment size. This is essentially the same as equation 2.42. A similar relation may be obtained through the Weibull distribution:

$$\frac{N(r)}{N} = 1 - \exp\left[-\left(\frac{r}{r_0}\right)^n\right] \tag{2.44}$$

which again reduces to a form similar to equation 2.42.

As noted in sections 2.3 and 2.4, Londe (1973), Priest and Hudson (1976), Rocha (1974) and Snow (1969) among others have noted the negative exponential distribution of discontinuity frequencies in various rocks. Fractal dimensions can be defined for spacing and length data in the manner discussed above. Poulton *et al.* (1990) computed fractal dimensions from various sources. While it is difficult to draw firm conclusions, it is possible, for rocks formed under similar tectonic or depositional regimes, to correlate higher fractal dimensions with higher strengths. For rocks of similar strength, higher fractal dimensions are associated with greater disturbance or weathering.

Discontinuity frequency defines, in $(1 - D)$ terms, the block size of discontinuous rock masses. If the discontinuity frequency follows a negative exponential distribution, then so should the block size. If the scale invariance can be applied to rock breakage, then fracture frequency, block size and particle size should all show similar fractal dimensions for a given rock mass. Higher fractal dimensions should be indicative of smaller block sizes and smaller particle size after blasting. The fractal dimension of the rock mass should be a function of the strength, tectonic history and lithology of the rock mass. Correlations should also exist between fractal dimensions and strength, RQD and hydraulic conductivity, with higher fractal dimensions indicative of lower strength, lower RQD and higher hydraulic conductivity.

Potentially the fractal dimension is a useful rock mass index, and used with care it may be possible to define definitively the characteristics of a rock mass with much greater precision than is the case with existing indices.

Mojtabai *et al.* (1989) and Ghosh (1990) have used the same approach to mapping of exposed benches in a typical Arizona porphyry sulphide deposit at Cyprus Miami Mine. Mojtabai *et al.* (1989) used cell mapping and Ghosh (1990) photo-analysis to determine block sizes. Figure 2.10 shows trace maps from three photographs of bench discontinuities by Ghosh (1990) and Figure 2.11 plots $N(r)$ against r as in equation (2.41).

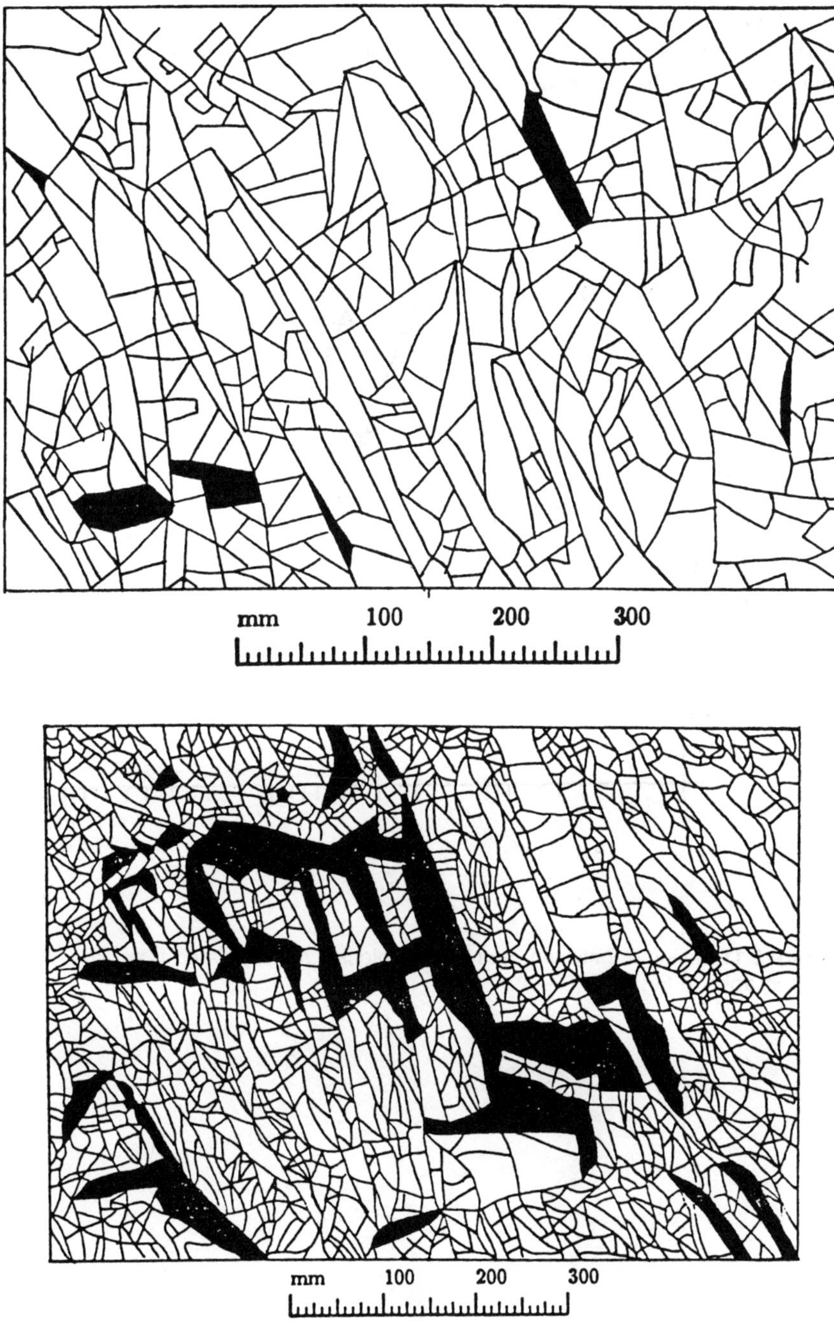

Figure 2.10 Joint trace maps developed from photographs of exposures in Schist (after Ghosh, 1990).

Although it is difficult to show how conclusions of fractal behaviour can be used in analysis, they demonstrate a uniformity in rock structure features, which may have considerable significance in estimating flow.

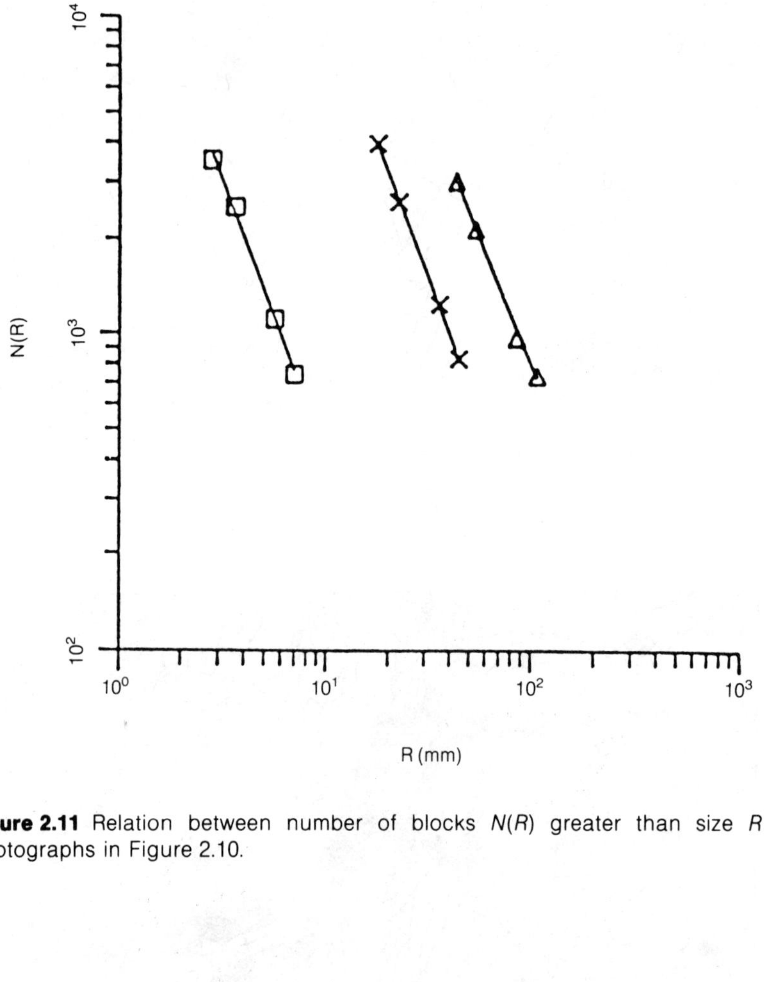

Figure 2.11 Relation between number of blocks $N(R)$ greater than size R for photographs in Figure 2.10.

Fluid flow in a single discontinuity 3

Flow through a single discontinuity is usually expressed in terms of the cubic law (equation 2.30) which is deduced from assumptions of laminar flow through apertures bounded by parallel plates. The cubic law is valid for natural, rough, uneven discontinuities when they are open (Iwai, 1976; Witherspoon *et al.*, 1980). However, in nature, they are rarely completely open, and flow, even through a single discontinuity, is analytically complex. One reason for this is the stress situation in the rock.

Flow through a single fracture or discontinuity in a rock mass is affected by both the effective normal stress acting normal to the plane of the discontinuity and its aperture. Since all natural discontinuities are subjected to effective normal stresses, then at some point the walls must be in contact. The contact area between the walls is a function of the discontinuity surface roughness (Gale, 1987). Thus, increases in contact area and resultant decreases in aperture resulting from geostatic stresses will restrict flow through the non-contact area of the discontinuity. A number of researchers (Gale and Raven, 1980; Gale, 1982, 1987; Bandis *et al.*, 1981) have studied the effect of discontinuity roughness, contact area and aperture on fluid flow with the objective of determining the fluid flow/deformation characteristics of various types of discontinuity under applied normal and shear stress conditions. It should be noted that most research has been carried out into changes in normal stress rather than shear stress.

3.1 DISCONTINUITY DEFORMATION – NORMAL STRESS

Normal loading tests have demonstrated that normal stress (σ_n)–normal deformation (ΔV_j) relations for a wide range of natural, unfilled joints are invariably non-linear, involving hysteresis behaviour during cycles of loading and unloading, and permanent deformation. Typical curves are shown in Figure 3.1.

The deformation characteristics of a discontinuity are affected (Bandis *et al.*, 1983) by various factors which include:

1. initial actual contact area, relative amplitude and vertical distribution of the aperture between the walls;

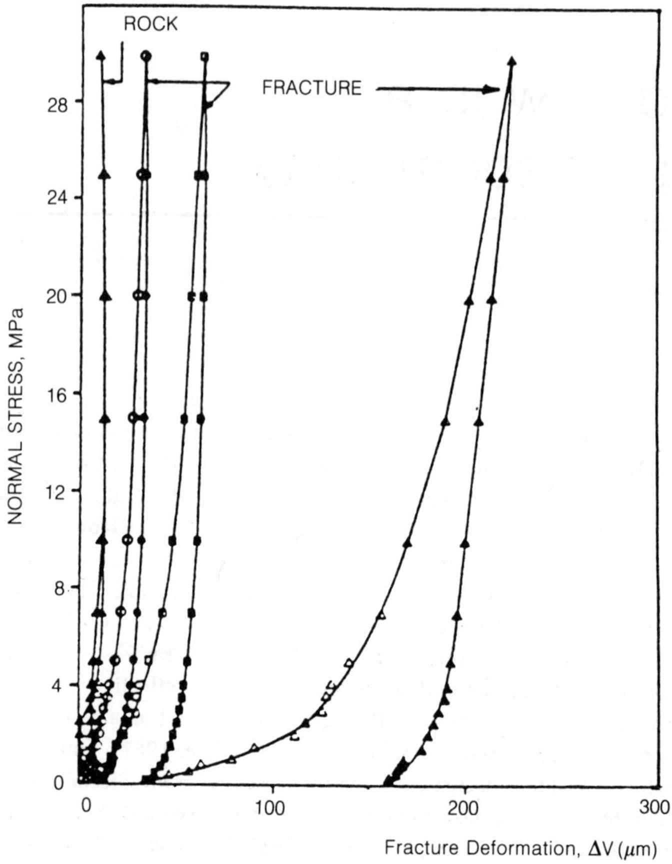

Figure 3.1 Discontinuity and rock deformation as a function of normal stress under cyclic loading (after Gale and Raven, 1980).

2. wall roughness;
3. strength and deformability of asperities;
4. thickness, type and physical properties of infilling material.

Two groups of models are generally used to predict the response of discontinuities to normal loading. The first group is based on the theory of elastic contacts and the Hertz theory (Brown and Scholtz, 1986; Swan, 1980, 1983; Johnson, 1983; Heuze, 1980). The apertures usually take the form of spherical segments, and the distribution of asperity heights is assumed from different probability density functions. The second group utilize empirical or experimental approaches. For example, Shehata (1971) proposed a logarithmic relation that states that the rate of discontinuity deformation is greater at lower normal stresses and that complete closure never occurs:

$$\Delta V_j = (\log \sigma_{n1} - \log \sigma_{n2})/C \quad (3.1)$$

where C is a function of the discontinuity stiffness, σ_n is the effective normal stress and ΔV_j is the closure.

Goodman (1976) proposed an empirical hyperbolic function to express the normal stress–deformation relationship:

$$\frac{\sigma_n - \xi}{\xi} = A\left(\frac{\Delta V_j}{V_m - \Delta V_j}\right)^t \tag{3.2}$$

where A and t are contants, ξ is the seating pressure, defining the initial condition for measuring the normal deformation, and V_m is maximum closure.

Bandis *et al.* (1983, 1985) proposed an empirical model for an interlocked joint:

$$\Delta\sigma = \frac{\Delta V_j}{c - b\Delta V_j} \tag{3.3}$$

where c and b are constants, and $c = 1/K_{ni}$ where K_{ni} is the initial stiffness, and $c/b = V_m$ is the maximum closure. These are given by empirical relations:

$$V_m = A + B(\text{JRC}) + C\left(\frac{\text{JCS}}{a_{t0}}\right)^D \tag{3.4}$$

$$K_{ni} = 0.02\left(\frac{\text{JCS}}{a_{t0}}\right) + 1.75\,\text{JRC} - 7.15 \tag{3.5}$$

where A, B, C and D are constants determined from cyclic loading tests, JCS is the joint roughness coefficient, JCS is the joint wall compression strength and a_{t0} is the initial mechanical aperture or initial true aperture.

Bandis *et al.* (1983) also proposed a relation for dislocated joints, given by:

$$\log \sigma_n = K_{ni} + M\Delta V_j \tag{3.6}$$

where M is a constant.

Barton *et al.* (1985) observed the contact area during normal loading tests. They proposed that when discontinuities were fully interlocked (i.e. unsheared), the ratio of total contact area (A_0) to total sample area (A) was related directly to a stress/strength ratio:

$$\frac{A_0}{A} \cong \frac{\sigma_n}{\text{JCS}}. \tag{3.7}$$

From tests, they showed that the ratio A_0/A was in the range of 0.4–0.7, depending on the normal stress magnitude. They also showed that in Iwai's (1976) tests, A_0/A was less than 0.001 when an effective normal stress level as low as 0.26 MPa was applied. At high normal stress levels (20 MPa) the ratio was in the range of 0.1–0.2.

3.2 DISCONTINUITY DEFORMATION – SHEAR STRESS

The shear stress–deformation characteristic of a rough discontinuity depends on the normal stress and the surface characteristics, since the shear deformation involves dilation and fracture of asperities and also variations in normal stress. Shear stiffness and peak shear displacement cannot be determined accurately. Theoretical analysis of shear deformation is, therefore, unsatisfactory and empirical criteria are used, as in normal deformation models. A hyperbolic function is frequently used to express the projected non-linear behaviour of sheared joints, especially in the pre-peak stress regime. For instance, Kulhawy (1975) suggests:

$$\tau = \frac{d_h}{m + nd_h} \tag{3.8}$$

where d_h is the shear displacement at shear stress τ, and m and n are constants.

Hungr and Coates (1978) defined the yield point of the $\tau - d_h$ curve. The basic form of their function was:

$$\tau = \frac{ut}{t - d_h} - u \quad \text{for } t < d_h \tag{3.9}$$

$$u = \frac{-0.9\, cf\sigma_n^2}{c\sigma_n - b} \quad \text{and} \quad t = \frac{-0.9 fb}{c(c\sigma_n - b)} \tag{3.10}$$

where u is the ratio of the yield (τ_y) to peak stress (σ_p), c is the ratio of the yield secant shear stiffness (K_{sy}) to normal stress, f is the peak friction coefficient under σ_n and b is the scale coefficient.

Barton *et al.* (1985) proposed that the peak shear displacement of a joint, d_p, be calculated from the empirical equation:

$$d_p = \frac{L_n}{500}\left(\frac{\text{JRC}}{L_n}\right)^{0.33} \tag{3.11}$$

where L_n is sample length and JRC is the roughness of this length of sample. In an interesting analytical study Bawden *et al.* (1980) showed that increasing the normal stress rapidly eliminates permeability increases associated with fracture dilatancy, and that beyond a normal stress limit dilatancy is completely overridden. Changes in fracture stiffness were also calculated to exert a strong influence on fracture permeability.

3.3 EFFECT OF STRESS ON PERMEABILITY

Actual data has been discussed in Chapter 1. With changing normal stress, the aperture, the relative contact surface area and the degree of asperity contact of a discontinuity are changed. Thus, flow of fluid through discontinuities and their hydraulic conductivities are directly related to stress and resultant deformation.

Gangi (1978) developed an asperity model, called a 'bed of nails', to determine the stress dependent hydraulic conductivity of fractured sandstone. In this model, the permeability variation of an individual discontinuity in the rock, subject to confining stress, σ_0, was ascribed to the asperities on the fracture in the form:

$$K(\sigma_0') = K_0\left[1 - \left(\frac{\sigma_0'}{\sigma_M}\right)^m\right]^3 \tag{3.12}$$

where σ_M is the effective modulus of the asperities within the discontinuity and is of the order of 1/10 to 1/100 of the asperity material bulk modulus, m is a constant $(0 < m < 1)$ which characterizes the distribution function of the asperity heights, σ_0' is the effective confining stress and K_0 is the zero pressure conductivity.

Gangi found that the effects of cyclic loading on permeability could be expressed in terms of an increase in m values. He also demonstrated that conical or wedge-like asperities could be treated as a 'bed of nails' model. However, Tsang and Witherspoon (1981) encountered some difficulties with the model due to a discrepancy in contact area at the same maximum stress when the 'bed of nails' model was applied to both the flow data and the stress–strain measurements on a granite fracture.

Walsh (1981) derived an analytical expression describing the variation of conductivity as a function of confining stress, σ_0, and pore fluid pressure, u_w:

$$\frac{\langle K\rangle}{\langle K_0\rangle} = \left[1 - \sqrt{2}\left(\frac{h}{a_{02}}\right)\ln\left(\frac{\sigma_0}{\sigma_p}\right)\right]^3\left[\frac{1 - b(\sigma_0 - u_w)}{1 + b(\sigma_0 - u_w)}\right] \tag{3.13}$$

where $\langle K_0\rangle$ and a_{02} are respectively hydraulic conductivity and half aperture at some reference stress σ_p; $b = \sqrt{[3(f/h)/E(1 - v^2)]}$, h is the r.m.s. value of the height distribution of the surface, f is the auto-correlation distance and E and v are Young's modulus and Poisson's ratio.

Walsh's formula was simplified by Bernabe (1986) to:

$$\langle K\rangle = (1 - A\log\sigma_0)^3\left(\frac{B - B'\sigma_0}{B + B'\sigma_0}\right) \tag{3.14}$$

where A, B and B' are constants depending on the geometry of the discontinuity and the elastic modulus of the rock type. The first, cubic, term of equations 3.13 and 3.14 represents the aperture effect; the second term is a tortuosity effect.

Walsh (1981) concluded that if the tortuosity term were equal to unity, the conductivity would vary with effective stress, σ_0' according to a relation:

$$K^{1/3} \propto A - B\ln\sigma_0'. \tag{3.15}$$

He also observed that the effective stress σ_0' was related to pore pressure u_w and confining pressure σ_c by an expression:

$$\sigma_0' = \sigma_c - Su_w \tag{3.16}$$

where S is about 0.9 for discontinuities with polished surfaces and 0.56 for tension fractures. S is dependent on how the volume and shape of the flow passage change with confining pressure when pore pressure is zero. Bernabe (1986) also showed that values of S were between 0.6 and 0.7 for granite. He concluded that S was strongly stress history dependent, and there was a decrease in S with increasing confining pressure.

Walsh and Brace (1984) showed that conductivity, elevated to a certain power n $(0 < n < 1/3)$ should be proportional to the log of σ_c:

$$K^n = (A \log \sigma_c + B) \tag{3.17}$$

where A and B are constants.

Swan (1983) derived a modified half-joint model. The relation between normal loading and hydraulic conductivity in slate joints being given by the simplified form:

$$\sqrt{\left(\frac{K}{K_0}\right)} = C - M \ln \sigma_\mathrm{n} \tag{3.18}$$

where C and M are constants and σ_n is the normal stress.

Gale (1982) demonstrated a relation between discontinuity conductivity (K_f) and stress for natural or induced fractures in gneissic granite in the form:

$$K_\mathrm{f} = \beta \sigma_\mathrm{n}^{-\alpha} \tag{3.19}$$

where β is the value of K_f (m/s) at a normal stress (σ_n) of 1.0 MPa. In his experiment, β varied between 2.81×10^{-3} and 4.28×10^{-4} m/s, and α varied between 0.129 and 0.276 for induced fractures. For natural fractures β varied between 7.54×10^{-3} and 7.38×10^{-4} m/s, α varied between 0.069 and 0.162.

Raven and Gale (1985) proposed a similar power law relation between the discontinuity transmissivity (T_f) and normal stress (σ_n) in the form:

$$T_\mathrm{f} = \beta \sigma_\mathrm{n}^{-\alpha} \tag{3.20}$$

where β(m^2/s) is the value of T_f(m^2/s) at σ_n equal to 1.0 MPa. α and β decrease with successive loading cycles, and tend to decrease with increasing sample size.

Tsang and Witherspoon (1983) studied the relation between flow rate and applied normal stress for mated and unmated joints. They indicated that when normal stress was applied to a joint sample, the resultant 'soft' mechanical behaviour gave rise to a sharp drop in flow under initial normal stress. Hence the more well-mated the discontinuity, the sharper the fall of flow with stress at a low level of applied normal load (Figure 3.2). This implies that the conductivity decreases more rapidly for well-mated joints than for unmated joints during initial loading.

Bandis *et al.* (1985) proposed that for larger aperture 'open' joints, the largest reduction in permeability would be predicted for smooth joints (lower JRC) and the least reduction for very rough joints (higher JRC) if the same initial aperture and increasing normal stress were assumed. For 'tight' joints, this characteristic may be reversed, especially for fairly planar joints.

Ryan (1987) indicated that closure was inelastic and time dependent. This reduced the volumetric flow rate and decreased the value of aperture calculated by the cubic law. In their results they indicated that over a period of two weeks, the aperture of the discontinuity decreased by about 7%, with a 3% decrease occurring during the first 30 minutes.

The effect of cyclic loading on the final conductivity of a discontinuity can be considerable. The permanent deformation of discontinuities will result in a relative

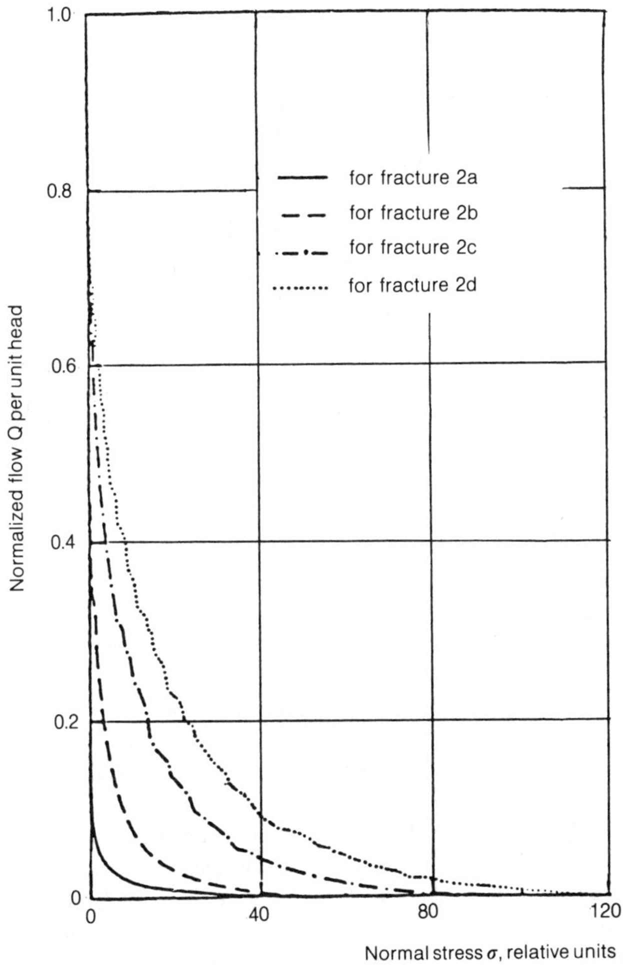

Figure 3.2 Fluid flow through discontinuities ranging from well-mated (fracture 2a) to unmated (fracture 2d). After Tsang and Witherspoon (1983).

decrease of permeability for each run at maximum closure. This can be surmised from Figure 3.1.

Gale (1982) and Raven and Gale (1985) showed that at a given normal stress over two or more loading and unloading cycles, flow rates and changes in flow rates also decrease. Permanent non-linear discontinuity deformation with distinct hysteresis occurs during loading and unloading cycles.

Bandis *et al.* (1985) showed that for a low stiffness joint a continuing reduction in permeability occurs under repeated loading with significant hysteresis. However, a high stiffness joint appeared slightly more conductive at the end of the second cycle than when first loaded, with virtually elastic recovery at very low stress levels. This implied that the stiffness of the joint after the first cycle was increased to a level preventing significant further closure, and resulting in a relatively larger conducting aperture at the end of the second loading.

In direct **shear** tests, because of the overall dilation of the discontinuity, the magnitude of the change of permeability will depend on the rock type, discontinuity aperture, roughness, normal stress, displacement history and the gouge within the joint.

Sharp and Maini (1972) recorded the permeability of a joint sample at various shear displacements and negligible normal stress. Changes in permeability of over two orders of magnitude were observed for a shear displacement of some 7 mm and an overall dilation of only 2 mm.

Makurat (1985) showed that the permeability of a natural rough joint increased by two to three orders of magnitude under a shear displacement of only 1 mm.

Tsang and Witherspoon (1983) indicated that under zero loading deformation, flow should first increase, reaching a maximum when the joint mismatch between upper and lower joints was about 1/2 of the large-scale undulation wavelength, and then decrease with further shear displacement.

3.4 EFFECTS OF SIZE

It is difficult to draw firm conclusions on the relation between the size of a discontinuity and its flow properties because of the different flow and boundary conditions used in various published papers (Figure 3.3). Some information can be obtained from analysis. The flow properties affected by size include aperture distribution, asperity height distribution (i.e. roughness) and the wavelength of the roughness undulation.

Tsang and Witherspoon (1983) showed that the large-scale roughness undulation of the discontinuity walls and the degree of mismatching between the two walls controlled the shape of the normal stress–displacement curve which illustrated the dependence of flow on normal stress. They suggested that the large-scale undulation wavelength of a discontinuity wall may be the key to approximating REV (Representative Elementary Volume) for stress-flow behaviour of single fractures. When the stress-flow behaviour of a fracture is investigated, the size of the rock sample should be larger than the typical wavelength of the roughness undulation.

Barton and Bandis (1980) proposed an empirical equation for Joint Roughness Coefficient (JRC) and Joint Wall Compression Strength (JCS) which included a scale effect:

$$\mathrm{JRC} = \mathrm{JRC_o}\left(\frac{L}{L_o}\right)^{-0.02\mathrm{JRC_o}} \quad \text{and} \quad \mathrm{JCS} = \mathrm{JCS_o}\left(\frac{L}{L_o}\right)^{-0.03\mathrm{JCS_o}} \tag{3.21}$$

where L is the sample length and subscript 'o' denotes the initial condition.

Neuzil and Tracy (1981) using a log-normal distribution of discontinuity aperture suggested that when specimen size is decreased, few of the larger least-frequent flow channels would be included because of truncation of the aperture frequency distribution. Thus small discontinuities should generally have smaller conductivity

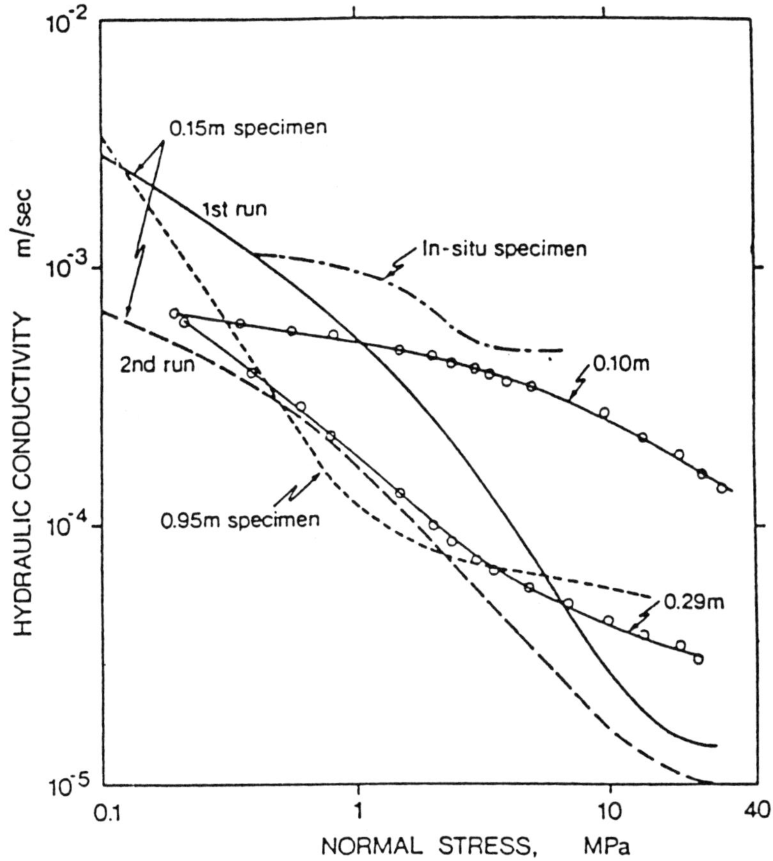

Figure 3.3 Effect of specimen size on the relation between hydraulic conductivity and normal stress (after Gale et al., 1982).

than larger specimens under the same applied normal stress. Similarly, Swan (1983) used a statistical argument based on asperity height distributions. He indicated that smaller discontinuities are more likely to have smaller apertures and stiffer deformation properties because they contain fewer tall or peak height asperities.

Raven and Gale (1985) indicated that smaller discontinuity surfaces may be characterized by fewer high asperities unevenly distributed over the smaller area, while larger surfaces may be characterized by a greater number of more variable height asperities more evenly distributed over the area. Under applied normal stress, the larger discontinuities have more asperities in contact, lower contact point stresses and more tortuous flow channels. Thus, for larger discontinuities, there will be less permanent deformation, lower flow rate and deformation hysteresis between loading and unloading cycles and greater deviation from the cubic flow law. In their experimental results, the effects of size are the opposite of those predicted by Neuzil and Tracy (1981) and Swan (1983).

3.5 APERTURE AND CONDUCTING APERTURE

Since a true aperture is almost impossible to define in practical engineering terms, the concept of hydraulic aperture or **conducting aperture** is used. The true aperture of a discontinuity cannot be measured directly for two reasons. First, even when maximum normal stress (i.e. maximum closure state) is applied, the discontinuity will still conduct water through the residual aperture. Secondly, because of the effects of roughness and contact area, the net effect of stress on aperture dimension can only be seen through hydraulic testing. Thus analysis is only possible by measuring the conducting aperture and using this to represent the true aperture.

Many aperture distribution forms for discontinuities have been investigated (Swan, 1983; Gale *et al.*, 1985). The aperture distribution of discontinuities is generally described as log-normal. Neuzil and Tracy (1981) used a log-normal distribution and incorporated a variable aperture into the cubic relation for fracture flow. Snow (1969) also found that the aperture distribution was very close to log-normal. Gale *et al.* (1985) discussed the results of fluid flow studies where resin grout was injected into fractures. They concluded that the resin thickness was log-normally distributed, and that the mean resin thickness was greater than the aperture computed from the flow tests.

Witherspoon *et al.* (1980) and Gale (1982, 1985, 1987) have studied the relation between aperture and flow rate. They proposed that the effective flow aperture (a_0) consisted of two components: an unknown residual aperture (a_{res}) which exists at maximum normal stress and a measured maximum aperture (a_{max}) as shown in Figure 3.4:

$$a_0 = a_{res} + a_{max}; \tag{3.22}$$

a_{max} can be obtained from experimental data if it is assumed that the cubic law is valid at maximum stress. Gale and Raven (1980) used a non-linear least-squares method to calculate the residual aperture, assuming that the cubic law is not satisfied.

Barton *et al.* (1985) and Bandis *et al.* (1983, 1985) have shown that a hypothetical initial **conducting** aperture (a_0) can be defined from experimental evidence as:

$$a_0 = \mathrm{JRC}_0^{2.5} \Big/ \left(\frac{a_t}{a_0}\right)^2 \tag{3.23}$$

where JRC_o is the initial joint roughness coefficient and a_t is the true initial mechanical aperture in microns. They also proposed an empirical form for a 'disturbed' joint to obtain an initial *mechanical* aperture:

$$a_t = \frac{\mathrm{JRC}_o}{5}\left(0.2\frac{\sigma_{cf}}{\mathrm{JCS}_o} - 0.1\right) \tag{3.24}$$

where subscript 'o' denotes the initial condition and σ_{cf} is the unconfined rock compressive strength.

Swan (1983) studied slates and suggested an empirical relation for obtaining initial aperture: $a_o = 9A$, where A is normal stress divided by normal stiffness. In this

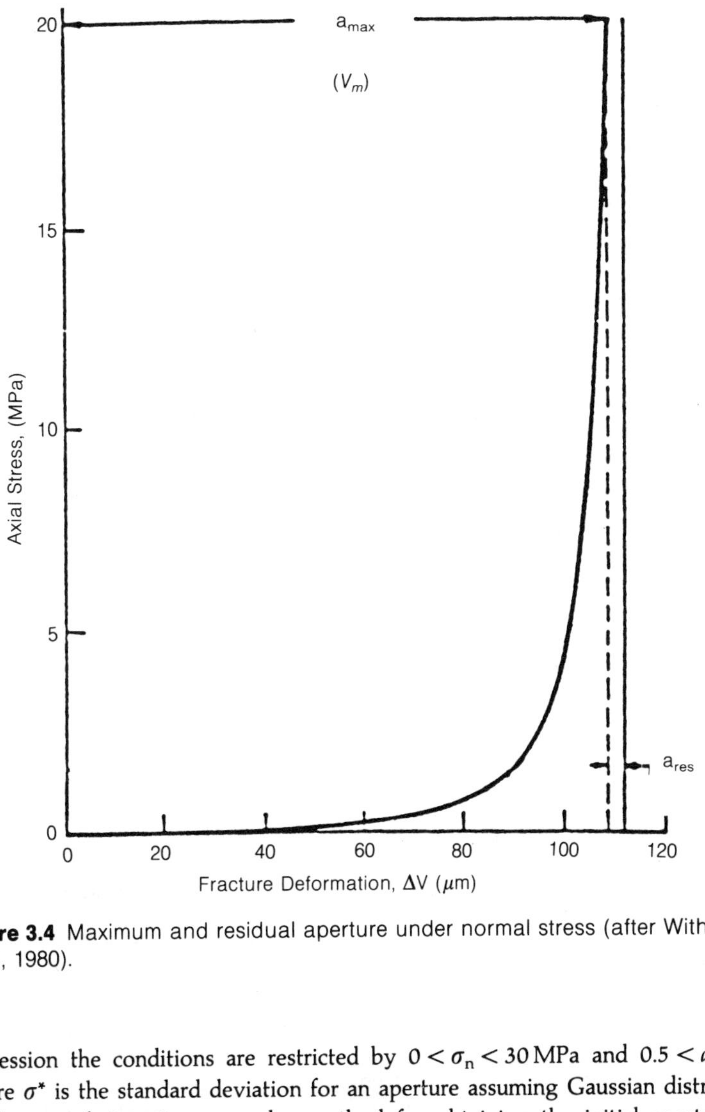

Figure 3.4 Maximum and residual aperture under normal stress (after Witherspoon *et al.*, 1980).

expression the conditions are restricted by $0 < \sigma_n < 30\,\text{MPa}$ and $0.5 < a/\sigma^* < 5$, where σ^* is the standard deviation for an aperture assuming Gaussian distribution.

Elliott *et al.* (1985) proposed a method for obtaining the initial aperture. This method assumes that for each effective normal stress, the aperture (a) can be calculated from flow data using the **cubic law** and a joint condition factor, F:

$$\frac{Q}{i} = \frac{C}{F} a_0^3 \quad \text{where } C = \frac{g}{12\mu}. \tag{3.25}$$

The aperture was then plotted against the measured closure (ΔV_j). They defined three conditions (Figure 3.5) for determining the initial aperture:

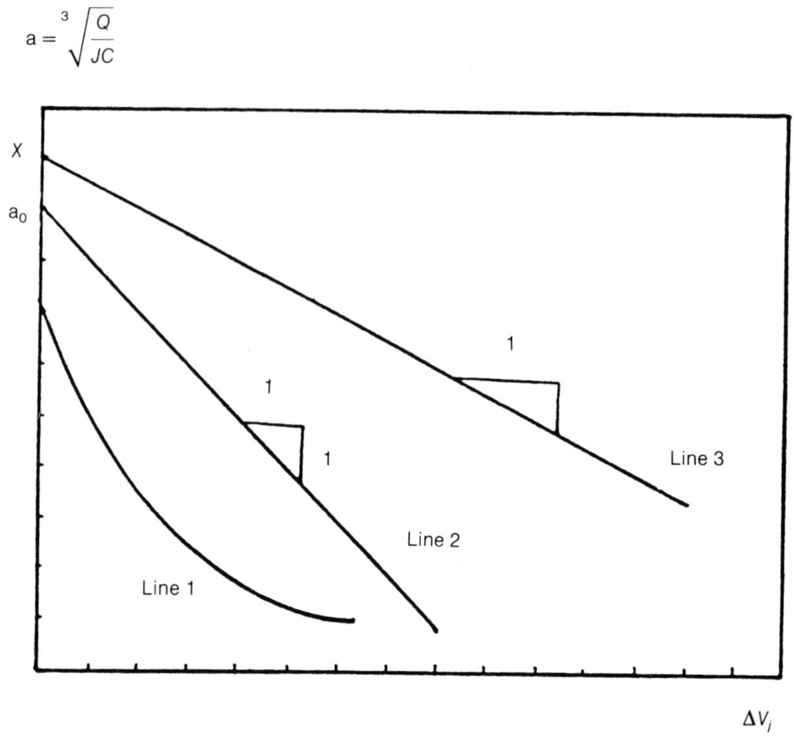

Figure 3.5 Relation between aperture, *a*, and measured closure (ΔV_j) for discontinuities (after Elliott *et al.*, 1985).

Table 3.1 Values of initial aperture and joint condition factor and aperture after one or more complete loading cycles

Data Souces	Sample	Joint state	Joint condition factor			Initial aperture (μm)		
			Cycle 1	2	3	1	2	3
Elliott *et al.* (1985)	2	Natural	1.26	–	–	169	–	–
	3		0.80	–	–	102	–	–
	4		13.3	2390	–	168	414	–
	5		1.03	199	–	86	205	–
Gale (1982)	41	Natural	74	167	–	252	221	–
	42		12	10	8	173	151	149
	40		0.8	40	–	205	187	–
	9	Induced	0.07	0.39	5.71	424	154	104
	10		11.8	63.3	–	497	195	–
	33		4.04	7.31	7.50	327	196	170
	34		30	81.1	–	283	269	–

1. If the flow rate per unit hydraulic gradient is not related to the aperture by a cubic function, the plot will be a curve (line 1). This implies that the method cannot be used since the cubic law fails.
2. If the plotted data are a straight line of unit gradient (line 2), then the cubic law is validated and the initial aperture is the value of the aperture axis intercept, a_0.
3. If the data describe a straight line which does not have unit gradient (line 3), then the cubic law is validated, but the initial aperture is calculated by $a = X/f$ where X is the value of aperture axis intercept and f is the slope.

Elliott *et al.* (1985) summarized the values of initial aperture from the joint condition factor, F. Values of their data and from their analysis of data published by Gale (1982) are shown in Table 3.1.

3.6 HYDRAULIC CHARACTERISTICS OF DISCONTINUITIES

Fluid flow tests on individual discontinuities have been performed to determine their hydraulic conductivity and to calculate their effective aperture. The relation between flow rate and hydraulic gradient has been deduced from theoretical and empirical approaches. Equations relating flow rate and hydraulic gradient are often assumed to be linear only for laminar flow at low velocities. At higher velocities they may become non-linear. The typical relation between flow rate and pressure is shown in Figure 3.6.

Linear laminar flow is assumed between parallel plates as saturated, laminar and incompressible. Then the transmissivity (T) or hydraulic conductivity (K) of a fracture is given by a cubic law similar to equation 3.25 if $F = 1$:

$$T = \frac{ga^3}{12\mu} \quad \text{or} \quad K = \frac{ga^2}{12\mu} \tag{3.26}$$

where a is the mean discontinuity aperture, μ is fluid kinematic viscosity and g is gravitational acceleration. These relations can be obtained experimentally or by use of the Navier–Stokes equation. Snow (1968b) proposed that if flow were laminar, the flow rate through a smooth parallel-walled channel of aperture a would be:

$$Q = \frac{\gamma a^3}{12\mu} Wi \tag{3.27}$$

where W is width and i is the hydraulic gradient.

Gangi (1978) proposed that the flow rate through a fracture in an impervious cylindrical rock sample of diameter D with a longitudinal fracture of length L and fracture permeability k_f would be:

$$k_f = \frac{a^2}{12}\frac{4a}{\pi D} = \frac{a^3}{3\pi D} \tag{3.28a}$$

$$Q = k_f \frac{\pi D^2 \Delta P}{4\mu L} = \frac{a^3 D}{12\mu}\frac{\Delta P}{L} \tag{3.28b}$$

where ΔP is the pressure difference over the fracture length.

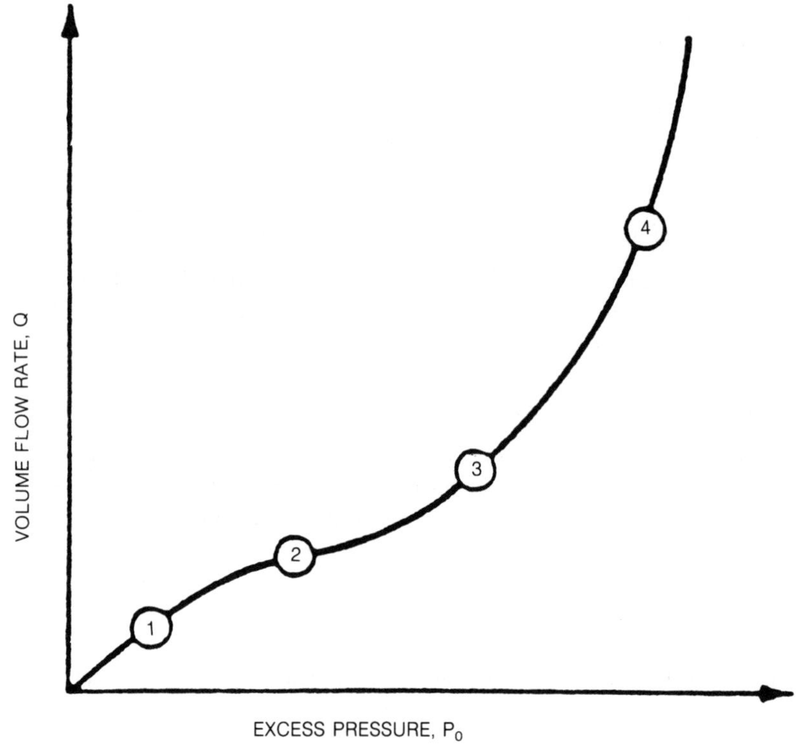

Zone 1 – Linear laminar regime
Zone 2 – Turbulence effects
Zone 3 – Turbulence offset by fissure expansion, or packer leakage
Zone 4 – Predominance of fissure expansion or packer leakage

Figure 3.6 Typical relation between flow rate in a fissure and pressure (after Louis, 1969).

There are two approaches to **non-linear flow** through individual discontinuities. One is an experimental approach based on the cubic law. The other is based on friction factor, Reynolds number and relative roughness. These terms are defined below.

Friction factor, λ, is related to driving head and flow velocity:

$$\lambda = \frac{2gai}{v^2} \tag{3.29}$$

where $v = Q/a$ is mean flow velocity, a is the average aperture, Q is the flow rate per unit width of the flow channel, i is the hydraulic gradient and g is the gravity acceleration.

Reynolds number, R_e, is expressed as an equivalent hydraulic diameter, D_h, defined as four times the average flow passage area divided by the perimeter. For a discontinuity, $D_h = 2a$ and $R_e = 2va/\mu$, where μ is kinematic viscosity. The critical Reynolds number commonly lies between 100 and 2300, increasing from the lower datum with decreased joint wall relative roughness.

Relative roughness, R_r/D_h, is the projection of the roughness (R_r) into the equivalent hydraulic diameter (D_h). R_r is the height of the surface asperities.

Louis (1969) has presented detailed empirical equations for computing discontinuity conductivity or flow rate corresponding to **laminar** and **turbulent** flow. Fractures containing infilling are assumed to have a permeability coefficent equal to that of the fill material.

Suitable values of R_r and the ranges of validity of laminar or turbulent regimes can be obtained. The conductivity and flow quantity (i.e. flow rate) are a function of aperture and fracture wall relative roughness (R_r/D_h) with a critical Reynolds number. The range of validity of these parameters related to relative roughness (R_r/D_h) and Reynolds number is shown in Figure 3.7.

Sharp and Maini (1972) proposed that if the effective hydraulic conductivity of a discontinuity is derived in terms of laminar flow, then the flow characteristics are described as either of the following:

1. linear laminar: $v = Ki$;
2. non-linear laminar: $v = K[i - B(i - i_{\lim})^n]$; (3.30)

where K is the effective hydraulic conductivity of the discontinuity, $i_{\lim}$ is the limiting gradient for linear laminar flow and B and n are empirically determined constants. Sharp and Maini (1972) also calculated the flow rate observed in their tests. They found that the flow rates were often not a function of the aperture to a power of three. This demonstrated that the cubic law was not generally applicable for rough discontinuities. Their results, expressed as $Q \propto a^n$, give exponent values:

	Rough discontinuity	Parallel plate
Linear laminar flow	$n = 2$	$n = 3$
Non-linear laminar	$1.2 < n < 2$	—
Fully turbulent flow	$n = 1.2$	$n = 1.5$

Maini *et al.* (1972) proposed that a non-linear flow rate/pressure relation might occur at lower hydraulic gradients. The form of non-linear law was characterized as:

$$v^n = C\frac{\partial P}{\partial X} \tag{3.31}$$

where v is mean velocity, n is the degree of non-linearity, P is the pressure, $\partial P/\partial X$ is hydraulic gradient and C is a constant depending on the viscosity of the fluid and

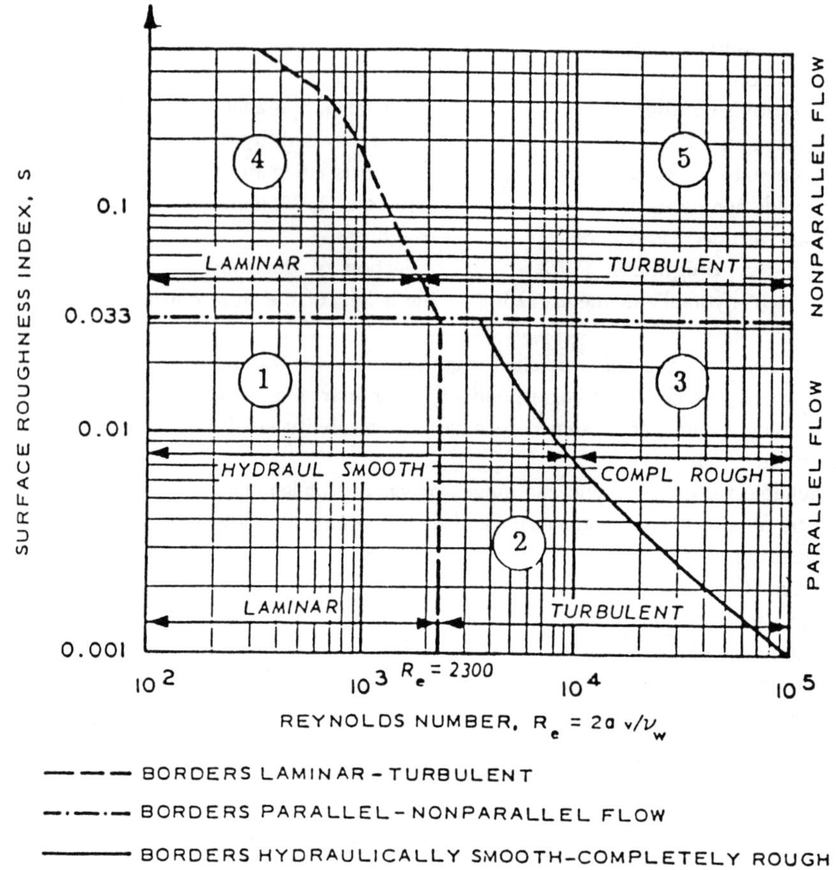

Figure 3.7 Range of validity of flow laws (after Louis, 1969).

the geometry of the discontinuity. For radial flow, this expression can become:

$$n \log Q + \log C = \log(P_0 - P_r) \tag{3.32}$$

where P_0 is the pressure at the borehole, P_r is the pressure at some distance r, Q is the flow rate and C is a constant which includes the hydraulic conductivity of the discontinuity, the viscosity of the fluid and the geometry of the borehole.

Elsworth and Goodman (1986) and Elsworth and Doe (1986) have mentioned that in conductivity testing of discontinuities, two flow laws relating flow velocity (v) to hydraulic gradient ($\nabla\phi$) during non-linear flow can be expressed as:

1. Forchheimer's Law using a polynomial expression:

$$\nabla\phi = bV + cV^2 \tag{3.33}$$

where b and c are constants and V is flow velocity in the direction of maximum gradient.

2. Missbach's Law using a power function:

$$\nabla\phi = Cv^m \tag{3.34}$$

where C is a proportionality constant and m is an exponent with magnitude ranging between 1 and 2.

3. Missbach's Law, rewritten as:

$$V = -K\Delta\phi^\alpha \tag{3.35}$$

where K is the hydraulic conductivity of the discontinuity and α is equal to unity for laminar and $\frac{1}{2}$ for turbulent flow in a rough-walled discontinuity.

Witherspoon *et al.* (1980) proposed that if flow was steady and isothermal, the flux per drop in head ($\Delta Q/\Delta h$) could be developed from Darcy's Law and be written in simplified cubic law form as:

$$\frac{\Delta Q}{\Delta h} = \frac{C}{f}a^3 \tag{3.36}$$

where c is a constant, equal to $\left[\dfrac{2\pi}{\ln(r_a/r_w)}\right]\left(\dfrac{\rho g}{12\mu}\right)$ for radial flow and $C = \left(\dfrac{W}{L}\right)\left(\dfrac{\rho g}{12\mu}\right)$ for straight flow and where $f \geqslant 1$ if $f = 1$ represents laminar flow.

Neuzil and Tracy (1981) proposed a model which led to a modified Poiseuille equation for flow including log-normal aperture distribution:

$$Q = \frac{\rho}{12\mu}Li[l]^3 \exp\left(3\xi + \frac{9\sigma^2}{2}\right) \tag{3.37}$$

$$v_{av} = \frac{\rho}{12\mu}i[l]^2 \exp(2\xi + 4\sigma^2) \tag{3.38}$$

where $[l]$ is the unit length and ξ and σ^2 are the mean and covariance of log-normal aperture distribution respectively. L is sample length normal to flow.

Tsang and Witherspoon (1981) have proposed a physical model where a single discontinuity was represented by a collection of voids and where closure resulted from a deformation of these voids. A weighted average aperture $<b>$ allowed flow rate to be expressed as a function of normal stress in two ways:

1. where flow is in a single fracture in an x, y plane with a pressure head maintained between $x = 0$ and $x = L$, then the flow rate is given by:

$$Q(x) = \frac{W}{L}\frac{\rho g}{12\mu} < b^3 >_x^{1/3} \frac{1}{< 1/b^2 >_y}[H(0) - H(x)] \tag{3.39}$$

where $<b^3>_x^{1/3}$ is an average over the width of the sample and $< 1/b^2 >_y$ is an average over the length of the sample in x and y directions respectively.

2. where divergent flow is from a centre borehole, radius r_a, radially into a cylindrical rock sample of radius r_w:

$$Q(r) = \frac{-2\pi}{f \ln(r_a/r_w)} \frac{\rho g}{12\mu} < b^3 >_r^{1/3} \frac{1}{< 1/b^2 >_\theta} [H(r_w) - H(r_a)] \qquad (3.40)$$

where $< b^3 >_r^{1/3}$ is the average of the variation of aperture transverse to the macroscopic radial flow, and $< 1/b^2 >_\theta$ is the average of the variation of the longitudinal to the radial flow direction. f is a numerical factor and dependent on r_w/r_a so that increasing r_w/r_a will decrease f. Tsang and Witherspoon (1981) indicated that if the outer radius is less than seven times the well bore radius, the prediction of radial flow is a poor approximation.

Thorpe *et al.* (1983) observed that the permeability of a 1 m diameter by a 2 m high specimen of jointed granitic rock might be attributed entirely to flow through discontinuities. The equivalent single discontinuity relating hydraulic conductivity (K_f) to flow rate (Q) was written as:

$$K_f = \left[\frac{(Q/\Delta h)^2 \ln(r_2/r_1)^2}{(12\mu/\rho g)(2\pi)^2} \right]^{1/3} \qquad (3.41)$$

where Q is the overall flow rate, Δh is the head, ρ is the fluid density, μ is dynamic viscosity, and r_1 and r_2 are the radii of the inner and outer flow boundaries. Thorpe *et al.* (1983) indicated that the measured closure of the major discontinuities was greater than the equivalent aperture change calculated by assuming flow through a single parallel-plate.

3.7 VALIDITY OF THE CUBIC LAW

The cubic law has been deduced theoretically from the assumption that fluid percolates through a smooth aperture under saturated, laminar, incompressible flow conditions. For a rough discontinuity, the cubic law may not be valid. Laboratory experiments on flow through rough discontinuities subjected to normal stress show evidence that significant deviations from the cubic law exist at comparatively high stress levels. For instance, Iwai (1976) showed that for rough-walled discontinuities under low normal stress, changes in aperture result in changes in flow rate consistent with the cubic law, while Gale and Raven (1980) indicated that the cubic law is not applicable to flow in deformable natural discontinuities where varying contact area contributes significantly to changes in flow rate. In their results, the values of the exponent are greater than three and increase with successive load cycles and increasing sample size.

Gale (1982) found that breakdown of the cubic law for discontinuity flow occurs in induced fractures subjected to normal stresses greater than 20 MN/m^2 with breakdown occurring at much lower stresses in natural discontinuities. He also suggested that fluid flow took a tortuous path when moving through a real discontinuity, thus deviating from the cubic law.

Tsang (1984) showed from his experimental results that when using the cubic law, there existed a 1–2 order of magnitude flow rate error resulting from neglecting the tortuosity. Tsang and Witherspoon (1981) stated that only at low applied stresses, when the discontinuities are essentially open, is the parallel-plate approximation adequate. As the contact area between the discontinuity surfaces increases with stress, the roughness of the discontinuity surfaces ceases to be a mere perturbation on a mean parallel-plate aperture. When the contact area of the surface is greater than 30%, the fluid flow rate can be two or more orders of magnitude smaller than predicted by a flow model which does not include tortuosity.

Gale *et al.* (1985) proposed that when the adjoining surfaces are not in contact, existing laboratory data supported the use of the cubic law, with appropriate corrections for roughness. Where surfaces were in intimate contact, corrections for roughness alone did not appear sufficient to evaluate the true fluid velocity.

Brown (1987) showed that the actual flow rate between rough surfaces is about 70–90% of that predicted by the parallel plate model (e.g. cubic law) for the range of joint closures expected during elastic deformation.

It may be concluded that the deformation behaviour of unfilled discontinuities is influenced by factors such as initial aperture, roughness, contact area, or wall strength, and that the deformation behaviour will influence the flow rate and hydraulic conductivity. The hydraulic conductivity of a discontinuity is strongly affected by changes in stress. As the contact area increases between the surfaces, the fluid flow rate will deviate from the cubic law. Thus it is necessary to consider the effect of tortuosity. At comparatively high applied normal stresses, discontinuities will also exhibit flow rate variations, due to the existence of residual aperture.

Flow varies from laminar to turbulent, depending on the discontinuity, roughness, flow velocity and hydraulic gradient. The validity of the cubic law is only satisfied for some conditions. For a tight and rough discontinuity the cubic law is not valid. In general, these relations can be predicted by using Forchheimer's Law or Missbach's Law.

Permeability and depth in rock masses

4

Fluid flow in a saturated rock mass is governed by the geometry and surface properties of intersecting networks of conducting discontinuities, whether or not the rock matrix is impermeable, and to a certain extent by depth. In the previous chapter the conductivity–closure behaviour of a single discontinuity was discussed, and it was seen that the hydraulic behaviour of a discontinuity aperture is affected by the effective normal stress across the fracture plane. In the case of a rock mass the hydraulic properties will therefore be affected by the geostatic stress. The next step in a study of fluid flow in a discontinuous rock mass is to see how flow through an idealized rock discontinuity system changes with depth, and how it relates to actual measurements.

4.1 HYDRAULIC PROPERTIES OF A DISCONTINUOUS MEDIUM

The **porosity** of a rock mass is defined as the volume of voids in the intact rock plus the volume of discontinuity openings divided by total volume of the rock mass. The porosity of intact rock can be obtained by laboratory tests. The porosity of discontinuities is more difficult to measure since they have variable apertures. Similarly, the **permeability** of a rock mass involves two parts: the permeability of the intact rock and of the rock mass discontinuity system. In the case of crystalline rocks, the permeability and porosity of intact rock are much less than those of the discontinuity system because the discontinuities have greater capacity for fluid flow than the rock matrix. In such cases, fracture permeability and fracture porosity are dominant.

To analyse the flow properties of a rock mass, the first and simplest approach is to assume that an equivalent medium involves a regular system of smooth planar fractures. Then the flow regime can be described by analogy to laminar flow between parallel plates. In this case, and assuming the validity of the cubic law, the hydraulic conductivity of a single discontinuity, parallel to its plane, is given by equation 1.8.

The hydraulic conductivity of a joint set is constant in every direction parallel to the set and its magnitude is given, from equations 1.8 and 1.9, by:

$$K = \frac{\gamma a^3}{12\, s\mu} + K_m \tag{4.1}$$

where s is the discontinuity spacing and a the aperture.

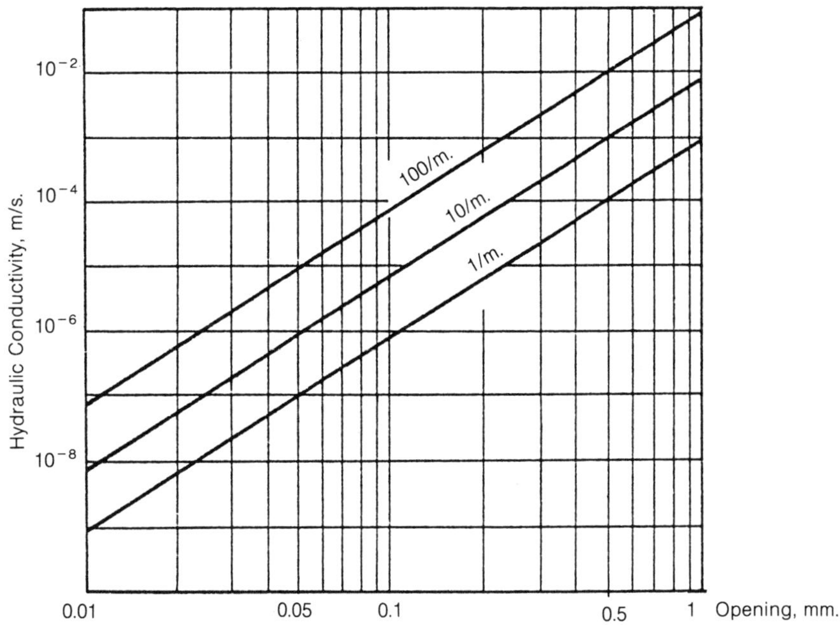

Figure 4.1 Hydraulic conductivity, frequency and conducting aperture for a rock mass containing parallel joint sets.

The hydraulic conductivity of a rock mass with three orthogonal joint sets with the same spacing and aperture is constant in three dimensional space, and is given by (Snow, 1968a):

$$K = \frac{2\gamma a^3}{12 s\mu} + K_m \tag{4.2}$$

In many circumstances, the hydraulic conductivity of a rock mass is several orders of magnitude higher than that of the intact rock. In this case, only discontinuity conductivity is important and the hydraulic conductivity of the intact rock can be neglected. In the case of impermeable intact rock, equations 4.1 and 4.2 can be plotted to relate hydraulic conductivity and aperture as shown in Figure 4.1. However, when the rock itself is porous or poorly jointed, or the joints are filled with material of low permeability, the above expressions will be invalid, and the hydraulic conductivity of intact rock must also be considered.

4.2 PERMEABILITY AS A FUNCTION OF STRESS

Empirical equations relating permeability to *in situ* stress have been considered in Chapter 1. An imposed stress will deform pores and discontinuities in a rock mass and may initiate new, preferentially orientated, fractures. Permeability of a rock mass is, therefore, fundamentally related to the stress field. The permeability of a number of rock types has been observed to decrease significantly with an increase in the mean or effective confining stress. Using a pressure transient technique, Brace *et al.* (1968), studied the permeability of Westerly Granite as a function of the effective confining pressure. They found that the permeability of the granite decreased from 350 ndarcies at 10 MPa pressure to 4 ndarcies at 400 MPa pressure (Figure 4.2), and noted that data obtained by Ohle (1951) from tests on a coarse-grained limestone suggestd a pressure effect not unlike the one they observed.

Nelson and Handin (1977) found that the matrix permeability of Navajo Sandstone decreased appreciably with increasing confining pressure. They normalized their data with respect to the permeability of the samples at 3.5 MPa confining pressure and observed decreases in permeability of up to 70% at a confining pressure of 69 MPa. The results of Lai (1971), Reynolds (1960) and Sutherland and Cave (1978) show that permeability of rock salt decreases as the mean confining stress increases. Relationships between permeability and effective confining pressure for different permeants are plotted in Figure 4.3. The curve labelled 'Argon' refers to samples of bedded salt, and the position of this curve between the ones for dome salt suggests that the geological origin of the salt is not a major factor. Lai also found that permeability increased with octahedral shear stress, τ_0. He summarized his findings in the empirical equation:

$$\log k = 0.62 - 0.00067\,\sigma_m + 0.00139 \log \tau_0 \qquad (4.3)$$

Note that in this equation k has units of cm^2, while both σ_m and τ_0 are measured in psi.

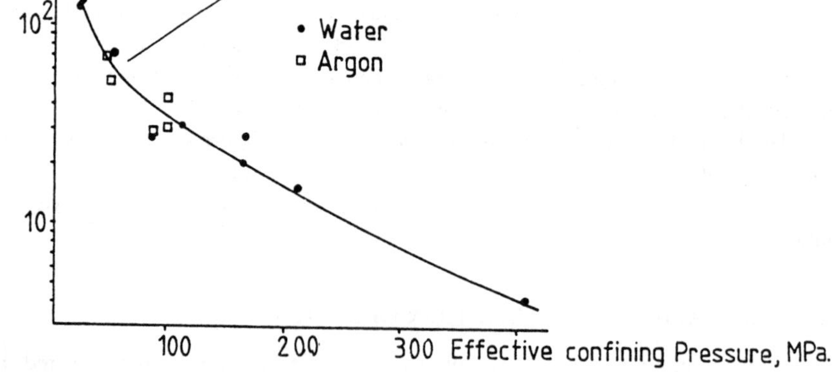

Figure 4.2 Variation of permeability with effective confining pressure (after Brace *et al.*, 1968).

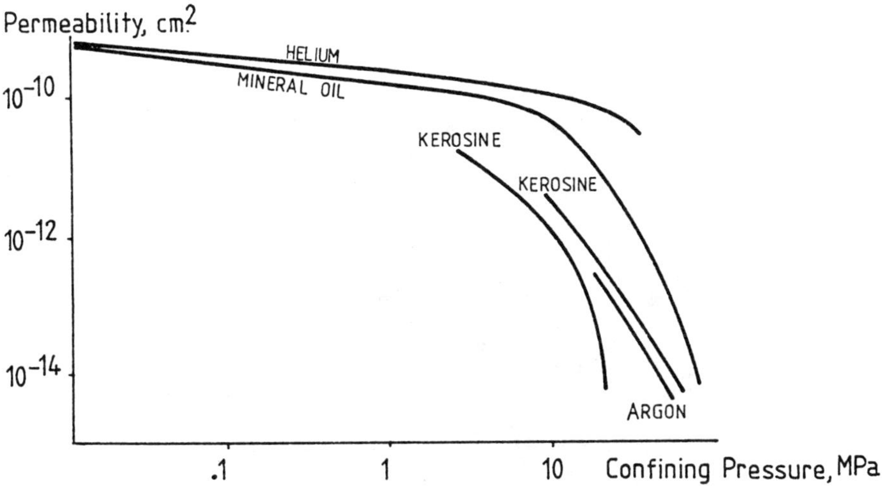

Figure 4.3 Permeability reduction with confining stress in rock salt specimens with various permeants (after Lai, 1971, Reynolds, 1960 and Sutherland and Cave, 1978). *Note*: 1 darcy is approximately equal to 10^{-8} cm^2.

The combined effects of τ_0 and σ_m on permeability can be summarized in a three-dimensional plot (Figure 4.4). Since Lai's results are typical of those presented in Figure 4.3, Figure 4.4 provides an interesting insight into the effects of the state of stress on the permeability of rock salt.

Zoback and Byerlee (1976) studied the permeability of Berea Sandstone and Ottawa Sand as a function of both confining pressure and fluid pressure. Although it has been suggested that permeability depends only upon effective stress (Brace *et al.*, 1968; Knutson and Buhor, 1963 and Vairogs *et al.*, 1971), Zoback and Byerlee (1976) observed that the effective stress law did not hold for sandstone and sand under either hydrostatic compression or triaxial stress. They showed that the permeability of Berea Sandstone under hydrostatic pressure was proportional to:

$$bu - \sigma \tag{4.4}$$

where b is a constant, u is the pore pressure and σ is the confining pressure and suggested the following empirical formula:

$$k(u, \sigma) = k_0 + c(bu - \sigma)k_0 \tag{4.5}$$

where k_0 is the intrinsic permeability. The constants c and b were found respectively to be 0.0143 and 4 mdarcies/100 kPa for flow normal to the bedding and 0.0598 and 2.2 mdarcies/100 kPa for flow parallel with the bedding. b approximates to unity for equilibrium and transport properties which obey the effective stress law.

The above work was based on laboratory tests, but Snow (1968c), Oda and Hatsuyama (1985) and Oda (1986) among others, have proposed and utilized empirical – usually logarithmic and exponential – functions relating rock mass permeability to

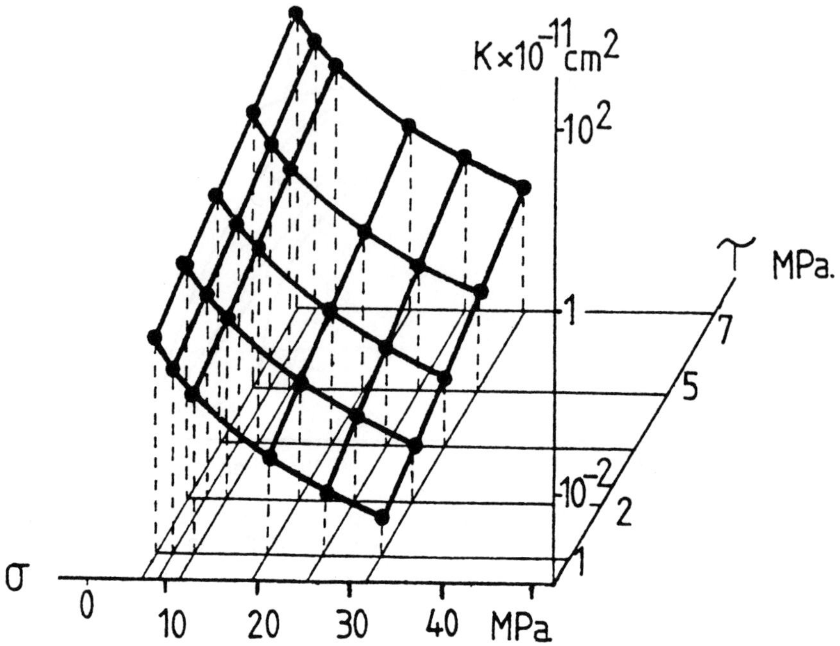

Figure 4.4 Relations between permeability, confining stress and octahedral shear stress (after Isherwood, 1979).

depth (section 1.6). However, there is strong evidence that permeability can vary by over four orders of magnitude at a particular site at the same depth (Figure 1.8). This suggests that zones of increased porosity and permeability can exist locally where localized clusters of high density and long fractures occur. The reason for this is related to the different hydraulic behaviour of rock masses with depth. These are usually characterized in hydrology by three or more zones, typically:

The **active zone**, which extends from the ground surface and includes the permeable overburden, the water table and the upper part of the bedrock. In many areas, the thickness of the active zone may be less than 50 m to 100 m. The extent of this zone usually depends on the depth of the zone of the weathering, the surface topography and the discontinuity characteristics of the rock. Significant systematic fracture permeability is typically greater than about 10^{-7} m/s. Singular features and fault systems with comparatively high permeabilities occur in this zone. Groundwater circulation with recharge and discharge areas, which are affected by evaporation, transpiration and precipitation, is relatively rapid and normally quite shallow. The level of the phreatic surface fluctuates due to variations in recharge, barometric pressure, water abstraction and stress fields. Typically it is in this zone that permeability varies with depth and stress field.

The **intermediate zone** is situated below the active zone. Singular features may extend into or through the intermediate zone, and thus create localized zones of higher permeability. The fracture system is probably fully saturated but porosity and storage are low. Systematic fracture or joint conductivities will generally by less than 10^{-7} m/s.

The **inactive zone** is a region of low hydraulic conductivity less than 10^{-8} m/s, and very low hydraulic gradients, less than 10^{-3} or 10^{-4}. Singular features may also extend into the inactive zone. Groundwater circulation is not rapid, and thus this zone has low hydraulic gradients, joint mineral infillings and stress closure. In igneous and metamorphic rocks, extensive zones of inactive groundwater flow may exist at depths as shallow as a few hundred metres or, at most, 500 m to 1000 m.

The presence of shales, clays and other impermeable beds within a stratification of low permeability, or formations whose permeability decreases significantly with depth may cause inactive zones to be located at shallower depths. In extreme cases the effects of weathering, tectonism, topography and interbedded aquifers and aquiludes may extend the active and intermediate flow zones to considerable depths.

4.3 JOINT SYSTEM MODELS

It is useful at this point to comment on joint system models. These are summarized by Dershowitz and Einstein (1988) and describe the joint system in a rock mass. The joint geometry is represented by a stochastic model because ideally each joint geometry parameter is characterized statistically. Since joint geometry does not always exist as a unique distribution, a large number of combinations may exist in the model. Although a number of models are available, only a relatively limited number can reasonably be used. The most suitable models are the orthogonal model, the Baecher disk model, the Veneziano model and the Dershowitz model. In each model some of the geometrical characteristics have specific relations with each other, while each can vary. Thus, when geometrical parameters can be predicted, joint system models can be represented as an idealized model of the rock mass geometry in a first rigorous stage. In each model joints are assumed as planar, and any joint location and autocorrelation processes are permissible.

4.3.1 THE ORTHOGONAL MODEL

Orthogonal models assume that the joints are arranged in two or three orthogonal sets of parallel joints (Figure 4.5(a) and (b)). The basic orthogonal joint system model comprises finite parallel joints with a constant space between the joints within each set. Another common orthogonal model assumes that joints in each set belong to a Poisson process which results in an exponential distribution of the spacings between joints. These joints are divided by bounded and unbounded joints (Dershowitz and Einstein, 1988). For the unbounded joints, joint shapes, sizes, and termination processes are not defined because the joint sizes are larger than the block volume size. For bounded joints, these parameters need to be defined. Joint terminations can be systematic at joint/plane intersections or independent of plane intersections. When joints are assumed to terminate at a joint/plane intersection, the joint shapes are

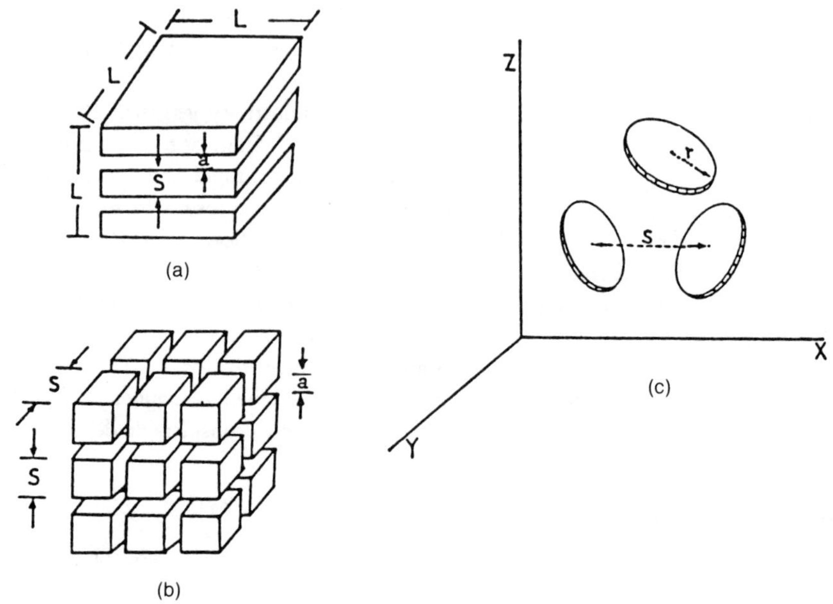

Figure 4.5 Three joint system models comprising (a) one parallel set of joints, (b) three orthogonal sets of fractures and (c) disc-type fractures.

rectangular and the joint size is defined by the spacing distribution between plane intersections. Although rectangles and squares of constant size are used, the model provides a reasonable approximation for many cases because rock blocks can be evaluated as rectangular prisms. Snow (1969) used this model to predict the hydraulic properties of a rock mass.

4.3.2 THE BAECHER DISK MODEL

The fundamental concept of the Baecher disk model (Baecher *et al.*, 1977) assumes that the joint shape is circular or elliptical. A joint set comprises parallel discs, the centres of which are located at random points in space (Figure 4.5c). The disc radii have a lognormal distribution. The joint centres within a volume are generated by a random variable with a Poisson distribution. This model has been extended by assuming that joint radii follow exponential and lognormal distributions, both of which produce a lognormal distribution of joint trace length (Dershowitz and Einstein, 1988). Joint locations can be determined through a regular pattern or a stochastic Poisson process, and can follow any orientation distribution or constant orientation. On the basis of this model, Long (1985) and Long and Witherspoon (1985) have developed an approach that includes any combination of joint size, location, and orientation assumptions. Long (1985) has also used this extended model to evaluate the permeability tensors assuming that joints are circular shapes.

4.3.3 THE VENEZIANO MODEL

The Veneziano model assumes that joint planes (not joints) are generated as Poisson planes which are located in space with a uniform distribution. These planes may have an orientation distribution. Then a Poisson line process on each joint plane is used to divide this plane into polygonal regions. These regions may belong to joints or intact rock. A certain proportion of these polygons is randomly marked as joints; the other polygons are defined as intact rock. This model leads to an exponential distribution of joint trace lengths (Dershowitz and Einstein, 1988). The disadvantage of this model is that the joint edges and joint intersections do not coincide because joints are marked in joint planes and the intersections between joints may be located in different joint planes.

4.3.4 THE DERSHOWITZ MODEL

The Dershowitz model is similar to the Veneziano model. The joint planes are generated by Poisson planes in space at uniformly distributed locations. The orientations may then follow a specific distribution. The intersections between these joint planes are used to define a series of lines on each joint plane. The lines divide each plane into polygonal regions and a certain proportion of the polygons on each joint plane may be defined as joints; the others are intact rock. The intersections of joint planes are defined as joint edges, so that all joint intersections occur at joint edges. The polyhedral faces of the joint blocks may be either intact rock or joints. The advantages of this model are that it can be defined as an orthogonal model for distinct rock blocks and have flexible orientation distributions such as polygonal joint shapes and polyhedral block shapes.

Joint system models are based on geometrical approximations of joint parameters and are used to characterize the joint geometry within a rock mass. Although these models cannot exactly represent the relation between joint geometry parameters within a rock mass, in practice, the stochastic natures of these models make it possible to approximately describe the natural spatial variation. For each model, some characteristics have a specific relation to the distribution form in each natural spatial variation. For instance, the Veneziano model leads to an exponential distribution of joint trace length; the Baecher model leads to a lognormal distribution; the orthogonal model leads to an exponential distribution of the spacing between joints.

In practice, joint geometries are not usually well characterized in a rock mass and in order to model them satisfactorily, it is necessary to compare the statistics of the model simulation with the sampled statistics and to continually update the model selection.

4.4 COUPLING, CLOSURE AND GEOMETRY

Many researchers have studied the relations between aperture closure and factors such as joint roughness, wall strength and stresses acting on the joint plane. Some analytical solutions used to define these relations are described in Chapter 3. However, the application of these solutions is rarely practical. The most satisfactory approach is through empirical relations and the best and most simple are those of Bandis *et al.* (1983, 1985).

Since the **mechanical aperture**, a_t, is often larger than the **conducting aperture**, a, rough-wall discontinuities usually have a larger aperture than equivalent smooth wall fractures, if both have the same conducting capacity (Bandis *et al.*, 1985). In order to study the relation between a_t and a, Bandis *et al.* (1983) proposed that the real mechanical aperture be related to a suitable roughness term (JCR) and to the conducting aperture which can be approximated by an empirical equation incorporating a suitable roughness term (equation 3.23), which provides an approximate model of hydromechanical coupling, valid for $a_t \geqslant a$:

$$a = (\text{JRC})^{2.5}(a_t/a)^{-2}$$

Figure 4.6 shows that when $a_t = a$, divergence increases with increases in JRC and with decreases in a within a range of a from 1 μm to 1000 μm. Bandis *et al.* (1983) show that $a = a_t$ (cubic law) may only be valid when apertures are very wide, or when the joints are exceptionally smooth. Bandis *et al.* (1985) also observed that divergence obeying the cubic law is modelled for wide apertures (> 1.0 mm) of exceptionally rough joints. No divergence is modelled for the smoothest joint unless conducting apertures are very small (< 0.01 mm).

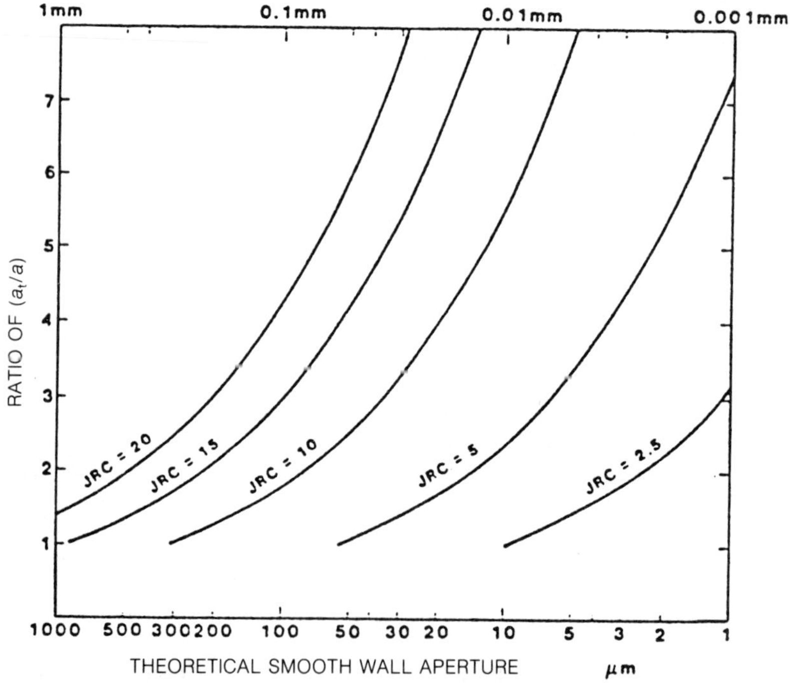

Figure 4.6 An empirical relation between mechanical aperture (a_t) and theoretical smooth wall or conducting aperture, a, in terms of joint roughness (JRC).

The required initial mechanical aperture (a_{t0}) can be estimated from equation 3.23 if the initial conducting aperture (a_0) and JRC are obtained by experiment. Another approximate method for solving for the initial mechanical aperture (a_{t0}) of a 'disturbed' sample under an assumed 'initial' normal stress, 0.001 MPa, can be obtained empirically (Bandis *et al.*, 1983, 1985), as in equation 3.24:

$$a_{t0} = \left(\frac{\text{JRC}}{5}\right)\left(\frac{0.2\sigma_{cf}}{\text{JCS}} - 0.1\right)$$

where σ_{cf}/JCS is dimensionless. The ratio of JCS/σ_c is usually in the range from 0.4 to 1.0 (Barton, 1976).

Normal closure is primarily dependent on the initial mechanical aperture, the roughness and the discontinuity wall strength. The loading and unloading deformation behaviour is approximated by a hyperbolic relation (equations 3.3 to 3.5). Constants A, B, C and D in equation 3.4 are dependent on the loading and unloading cycles which are a function of stress history. These values have been calculated by Bandis *et al.* (1983), using multiple regression methods for various rocks. If it is assumed that the first run cycle from their experimental data can be represented as *in situ* properties, values of these constants are $A = -0.2960$, $B = -0.0056$, $C = +2.2410$ and $D = -0.2450$.

In order to evaluate rock mass porosity and permeability, a **joint system model** must be used. The three simplest and most appropriate models are those in Figure 4.5, comprising one parallel set of fractures, three orthogonal sets and the Baecher disc model. For a parallel set of fractures, the rock mass is assumed to have the same spacings and conducting apertures. Rock mass or fracture permeability and porosity depend on the size of these openings. The idealization of discontinuities as a parallel set of fractures makes it possible to obtain a valid approximation of rock mass porosity and the permeability of the rock mass when the initial conducting aperture, spacing, JRC and JCS are estimated. The relation between permeability and porosity of a parallel set of fractures with spacing (s) is given by equation 4.1:

$$n = \frac{a}{s} \text{ and } K = \frac{\gamma a^3}{12\mu s}$$

For a cubic system of fractions, the rock mass is assumed to have the same spacing in three orthogonal directions. The permeability of three orthogonal sets is twice the permeability of one parallel set (Snow, 1968). Total porosity and permeability can thus be expressed as (equation 4.2)

$$n = \frac{3a}{s} \text{ and } K = \frac{2\gamma a^3}{12\mu s}$$

4.5 RELATIONS BETWEEN CONDUCTING APERTURE AND DEPTH

In order to evaluate the relation between conducting aperture and depth, using the empirical method suggested by Bandis *et al.* (1983, 1985), JRC, JCS, initial conducting aperture, a, and *in situ* stress must be known. Values of JRC, JCS and a can be estimated from field and laboratory investigations, if tests are carried out on undisturbed joint specimens in the latter case. The vertical *in situ* stress can be assumed from the weight of a vertical column of rock as $\sigma_v = 0.027\,z$(MPa) where z is vertical depth in metres. Thus, the empirical relations can be modified relatively easily to estimate fracture conducting aperture with depth, assuming no matrix flow, as follows:

Initial mechanical aperture is estimated from equation 3.23, given by:

$$a_{t0} = \sqrt[2.5]{(a_0\,\mathrm{JRC})} \tag{4.6}$$

Mechanical aperture in terms of depth is estimated from equation 3.3:

$$\begin{aligned} a_t &= a_{t0} - \Delta V_j = a_{t0} - \frac{\sigma_n V_m}{K_{ni} V_m + \sigma_n} \\ &= a_{t0} - \frac{0.027\,zV_m}{K_{ni} V_m + 0.027\,z} \end{aligned} \tag{4.7}$$

Conducting aperture in terms of depth is estimated from equation 3.23:

$$a = \frac{a_t^2}{\mathrm{JRC}^{2.5}} = \frac{1}{\mathrm{JRC}^{2.5}}\left(a_{t0} - \frac{0.027\,zV_m}{K_{ni} V_m + 0.027\,z}\right)^2 \tag{4.8}$$

Two extreme cases of an 'open' and a 'tight' discontinuity or fracture are used to illustrate (Lee and Farmer, 1990a,b) the effects of various values of JRC, JCS and a_0 on conducting apertures at depth in Figure 4.7. In the case of the open fracture

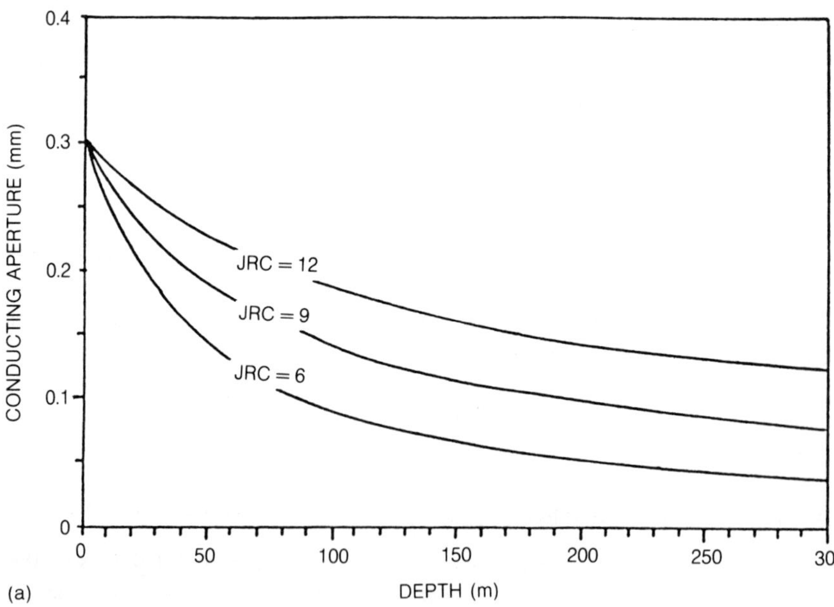

(a)

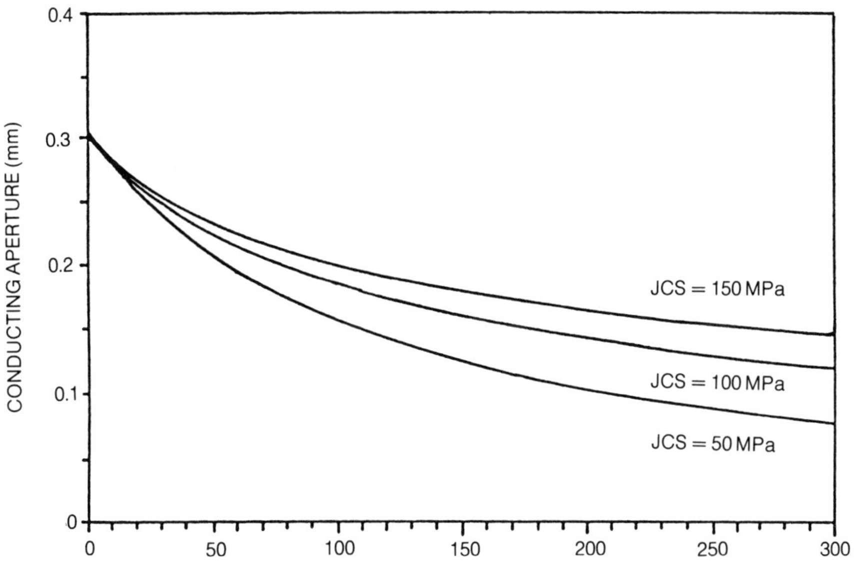

(b)

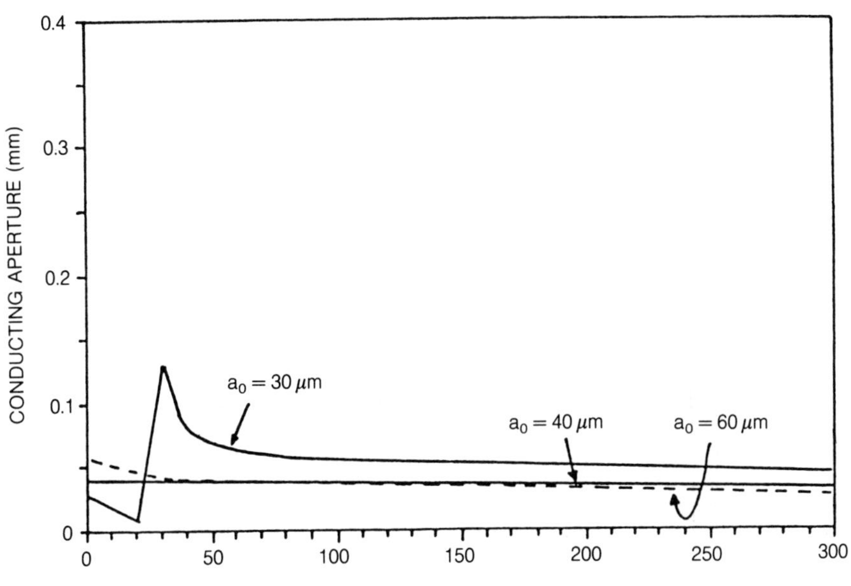

(c)

Figure 4.7 Effect of changes in JRC, JCS and a_0 on the relation between conducting aperture and depth: (a) with fixed JCS = 100 MN/m^2, $a_0 = 300\,\mu$m; (b) with fixed JRC = 12 and $a_0 = 300\,\mu$m; (c) with JCS = 125 MN/m^2 and JRC = 5.

an initial conducting operative $a_0 = 300\,\mu$m is assumed with a constant JCS = 100 MN/m^2. Changes in aperture with depth are illustrated in Figure 4.7a. The largest reduction in initial conducting aperture is predicted for a fairly planar discontinuity (JRC = 6) and the least for a moderately rough one (JRC = 12). With a constant JRC = 12, illustrated in Figure 4.7b, the largest reduction in conducting aperture is predicted for the weaker wall strength (JCS = 50 MN/m^2) and the least for the stronger wall strength (JCS = 150 MN/m^2).

In the case of the tight fracture, some interesting features are demonstrated by the conducting aperture–depth plot. Examples using JRC = 5.0, JCS = 125 MN/m^2 and $a_0 = 30, 40, 60\,\mu$m are shown in Figure 4.7c. It would be expected that the initial conducting aperture would change relatively little over a small depth range. However,

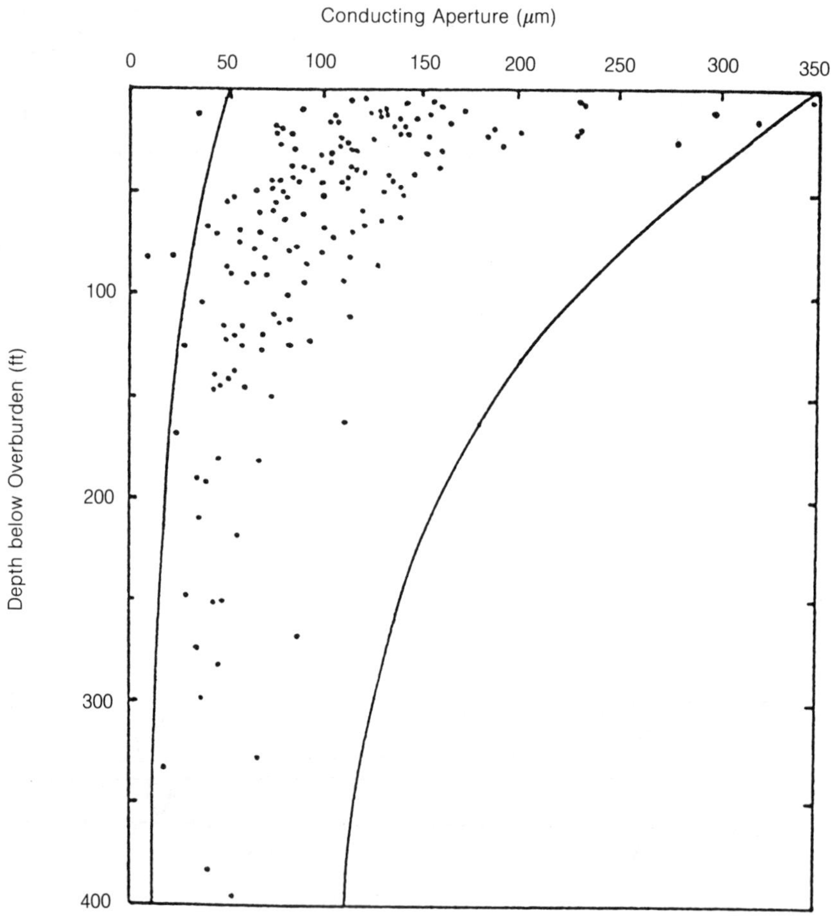

Figure 4.8 Comparison of field data from various dam sites (Snow 1968b) with estimates of aperture. The envelopes are based on $a_0 = 350\,\mu$m (RHS) and 50 μm (LHS), JCS = 95.8 MN/m^2, JRC = 7.4. *Note*: 1 ft = 0.305 m.

Table 4.1 Rock type, weathering state, JRC and JCS (data from Bandis *et al.*, 1983)

Rock type	Weathering state	JCS (MN/m^2)	JRC
Slate	F	175	4.0–5.0
	MW	122	5.3
	W	77	6.0–6.8
Dolerite	F–SW	167–182	6.1–6.8
	M	60–76	6.0–8.7
Limestone	F–SW	152–170	5.6–11.4
	MW	94–120	5.8–15.0
	W	35–50	5.0–16.0
Siltstone	F	105	8.0–1.8
	MW	42–67	6.1–7.5
Sandstone	F–SW	74–95	5.4–11.9
	MW	44–58	5.1–14.1
	W	22–25	5.0–6.0
		Average 95.8	Average 7.4

F: fresh joint; SW: slightly weathered joint; MW: moderately weathered joint; W: weathered joint.

for $a_0 = 30\ \mu m$ there was the highest maximum closure; and for $a_0 = 60\ \mu m$ the lowest closure. The reason for this may be that since JCS changes with increasing depth (i.e. increasing normal stress), the stiffness K_{ni} (equation 3.3) of the discontinuity increases to a level inhibiting significant fracture closure, then resulting in a relatively larger maximum closure state for the small conducting aperture.

As a result, when the open fracture involves a constant JRC, the largest reduction in conducting aperture is for weaker wall strength (lower JCS). When the fracture has a constant wall strength (constant JCS), the larger reduction is for the smoother fracture (lower JRC). However, for tight fractures, the conducting aperture decreases inconsistently with increasing stress since increasing stress may significantly change the value of JRC and JCS.

Snow (1968b) obtained detailed measurements of average fracture aperture against depth (Figure 1.4) in a seminal paper based on observations at various dam sites. Taking Snow's values for the maximum initial conducting aperture $a_{max} = 350\ \mu m$ and minimum initial conducting aperture $a_{min} = 50\ \mu m$ and taking the average value of JRC and JCS for a similar suite of rocks by Barton *et al.* (1985) as JRC = 7.4 and JCS = 95.8 MN/m^2 (Table 4.1), the curves of conducting aperture against depth are plotted in Figure 4.8. These show that the range of calculated values has good agreement with measured values.

4.6 ESTIMATION OF ROCK MASS PERMEABILITY BY EMPIRICAL METHODS

Illustration of methods of permeability estimation is constrained by lack of published data. Two of the better data sets are by Snow (1968b) and Tiren *et al.* (1985) although

Snow does not include permeability measurements. Fracture porosity data can, however, be determined by the same process and in this case is a valid comparison with Snow's porosity data.

Two of the models in Figure 4.5, those of one parallel set and three orthogonal sets of fractures, can be used to illustrate a simple empirical method of estimating rock mass permeability using equations 4.6–4.8, and to compare this with results obtained by Snow (1968b) and by Tiren *et al.* (1985).

The procedure for computing permeability or porosity is:

1. conducting aperture is obtained as in equation 4.8.
2. conducting aperture is substituted in equations 4.1 (parallel set) and 4.2 (orthogonal system) to determine permeability or porosity.

Input data, actual and estimated for six rocks types in Snow's case is summarized in Table 4.2.

The relevance of the approach is illustrated in Figure 4.9. The assumption of three orthogonal sets fits the data for granite, metavolcanics, gneiss and miscellaneous blocky rocks quite well. Slate and phyllite and shale and sandstone data fall between the models of one parallel set and three orthogonal sets. Thus it seems that porosity data for crystalline rocks may be simulated by the three orthogonal sets model; for sedimentary rocks, less accurately by both models.

Tiren *et al.* (1985) published data on rock mass permeability measurements in granodiorites with hornblende and biotite. The maximum spacing, s_{max}, and minimum spacing, s_{min}, are selected as $s_{max} = 0.5$ m and $s_{min} = 0.02$ m from borehole data in their paper. a_{max} is taken as 350 μm; a_{min} as 50 μm; JCS as 110 MN/m^2 and JRC as 8.

Figure 4.10 shows that there is a good fit using the three orthogonal sets assumption.

It may be concluded, therefore, that a model for rock mass porosity and permeability estimation based on three orthogonal discontinuity sets gives a good approximation

Table 4.2 Minimum spacing and maximum spacing for various rock types (data from Snow, 1968b) and estimates of JCS and JRC

Rock type	Minimum spacing ft (m)	Maximum spacing ft (m)	JCS (MN/m^2)	JRC
Granite	4 (1.22)	110 (33.53)	130	8
Metavolcanics	4 (1.22)	95 (28.90)	120	7
Sandstone and shale	6 (1.83)	20 (6.09)	60	6
Gneiss	5 (1.52)	45 (13.71)	100	7
Slate and phillite	4 (1.22)	25 (7.62)	70	6
Quartzite, serpentine, limestone and diabase	5 (5.24)	106 (32.31)	120	8

$\Delta_{max} = 350\ \mu$m, $\Delta_{min} = 50\ \mu$m

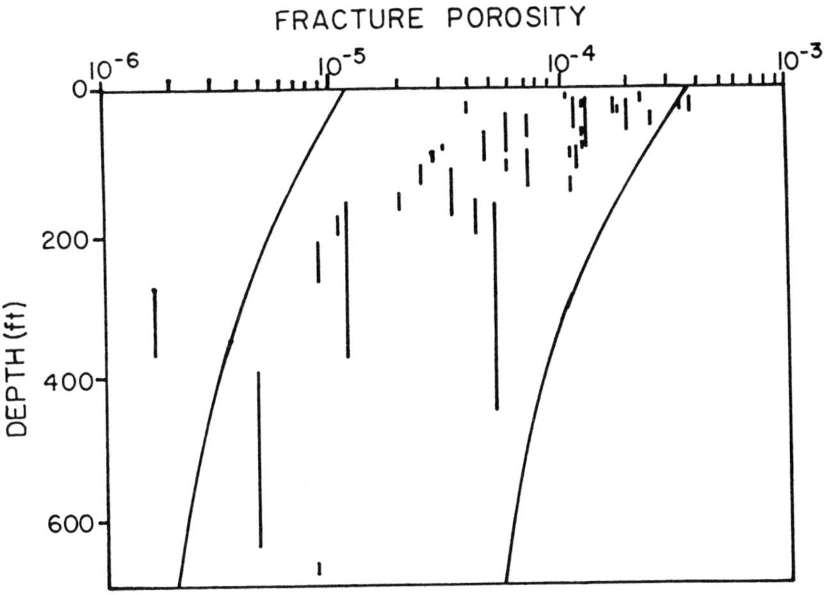

(a)

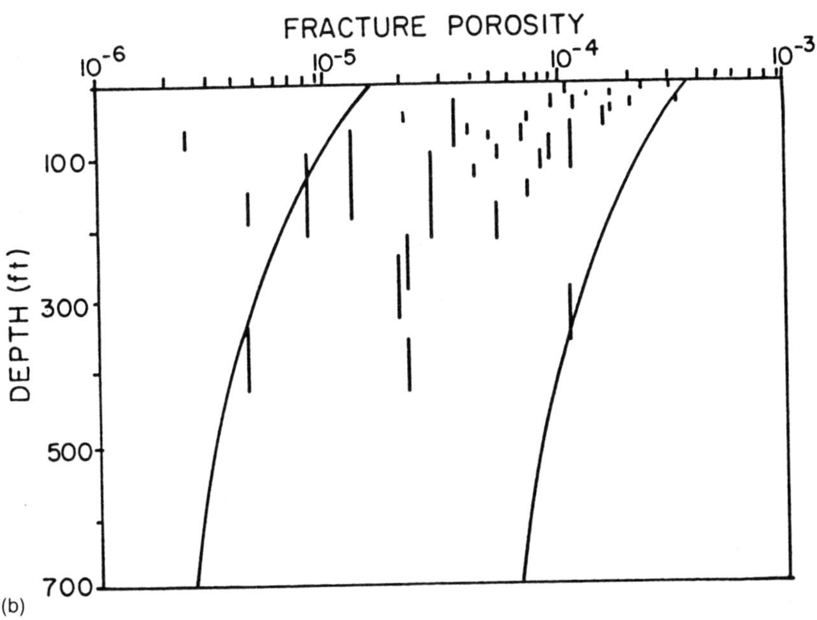

(b)

Figure 4.9 Measured fracture porosities and estimated envelope curves with depth from Snow's (1968b) data on (a) granites, (b) metavolcanics, (c) gneiss, (d) slates and phyllite, (e) sandstones and shales and (f) quartzites, serpentine, limestone and diabase. Note the solid lines represent an assumed orthogonal joint pattern; the hatched lines, parallel joints. *Note*: 1 ft = 0.305 m.

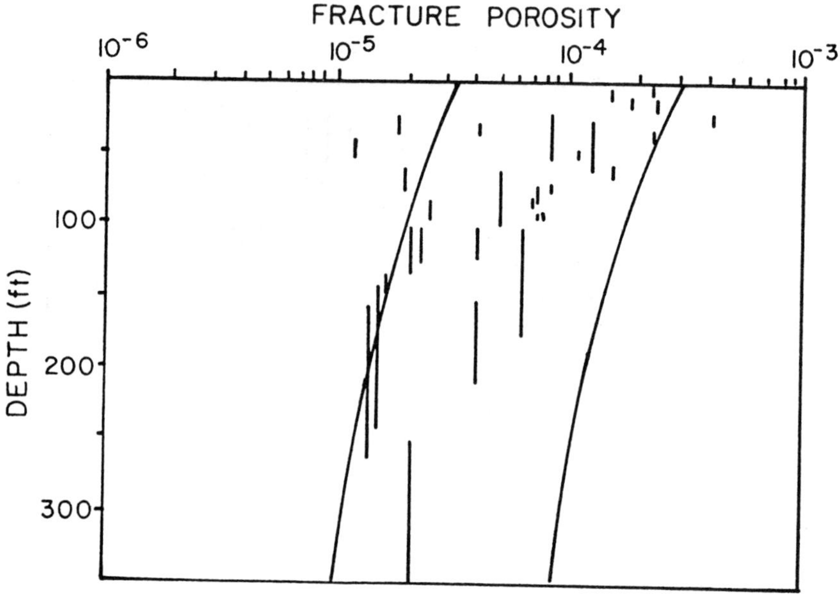

(c)

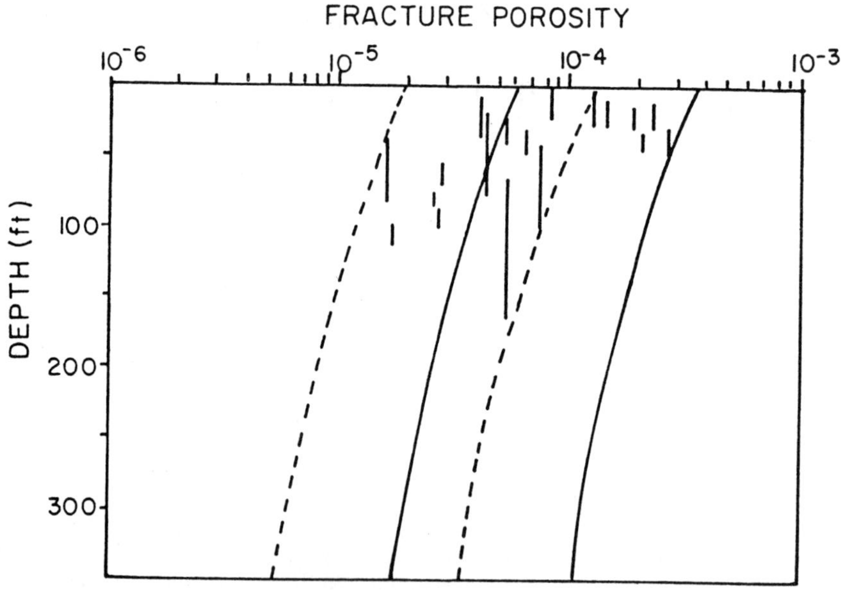

(d)

Figure 4.9 *contd.*

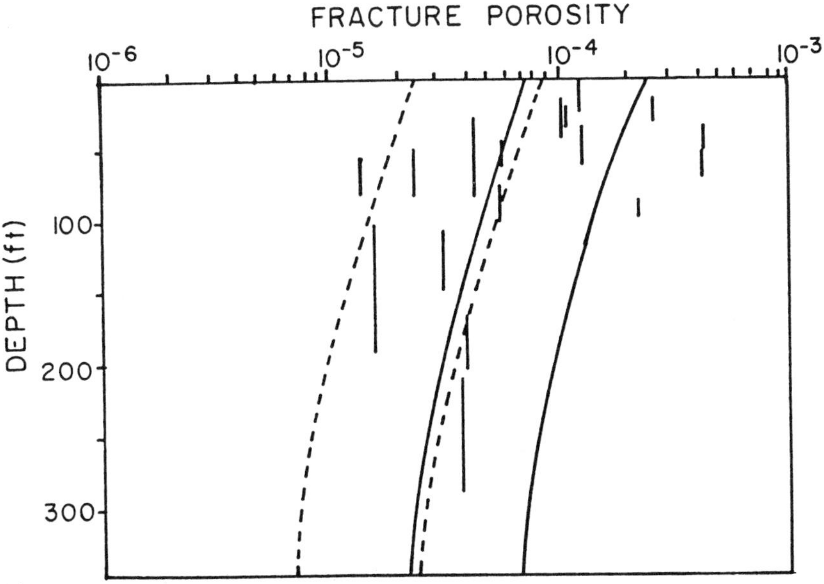

(e)

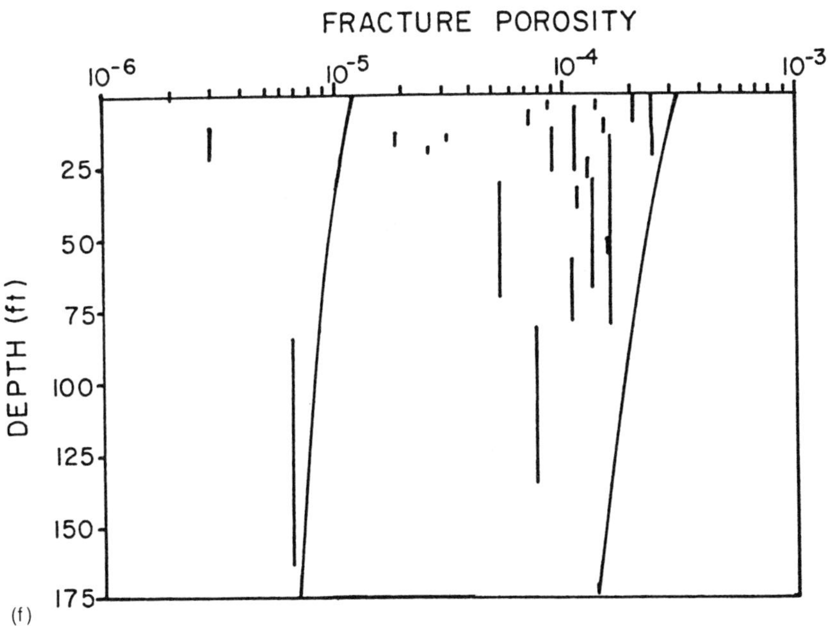

(f)

Figure 4.9 *contd.*

for crystalline rocks. For sedimentary and directional rocks, a model incorporating orthogonal and parallel sets of discontinuities can be used, but is less accurate.

Initial conducting apertures depend on many factors such as fracture surface characteristics, geostatic stress and geological and geometric characteristics. Usually the initial conducting aperture is obtained from laboratory flow tests under zero normal stress or evaluated from surfaces at zero depth. Elliott *et al.* (1985) have proposed a method from laboratory data with a joint condition factor to evaluate initial conducting aperture (section 2.8). This method is based on the cubic law, but by using this method, the initial conducting aperture may be evaluated quite accurately.

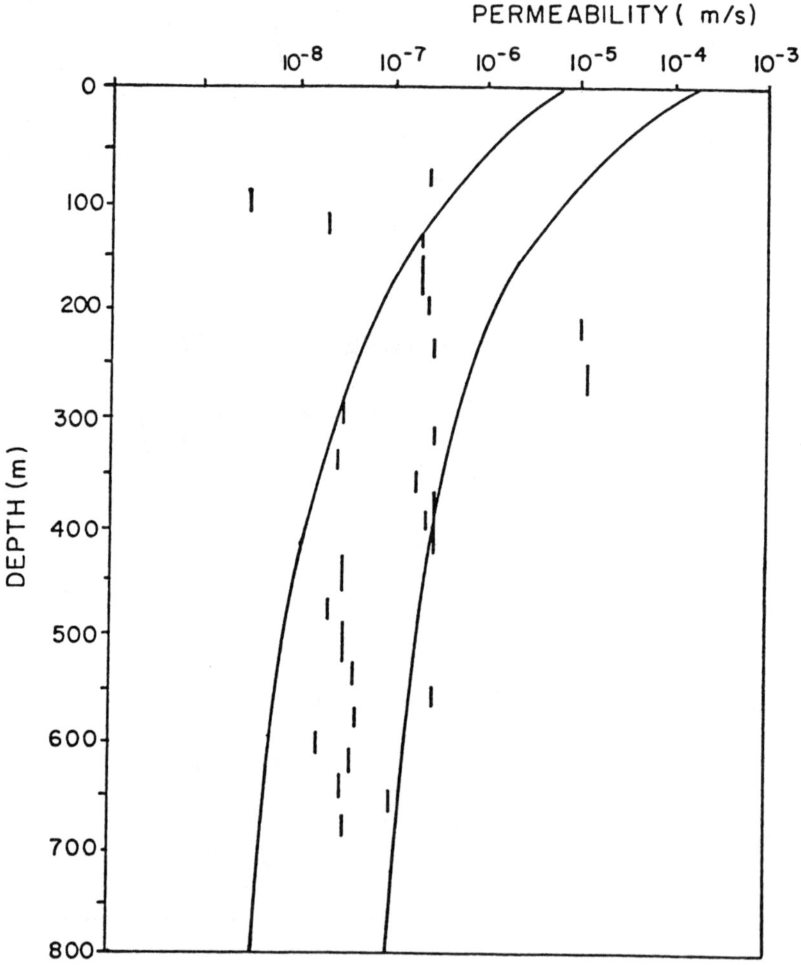

(a)

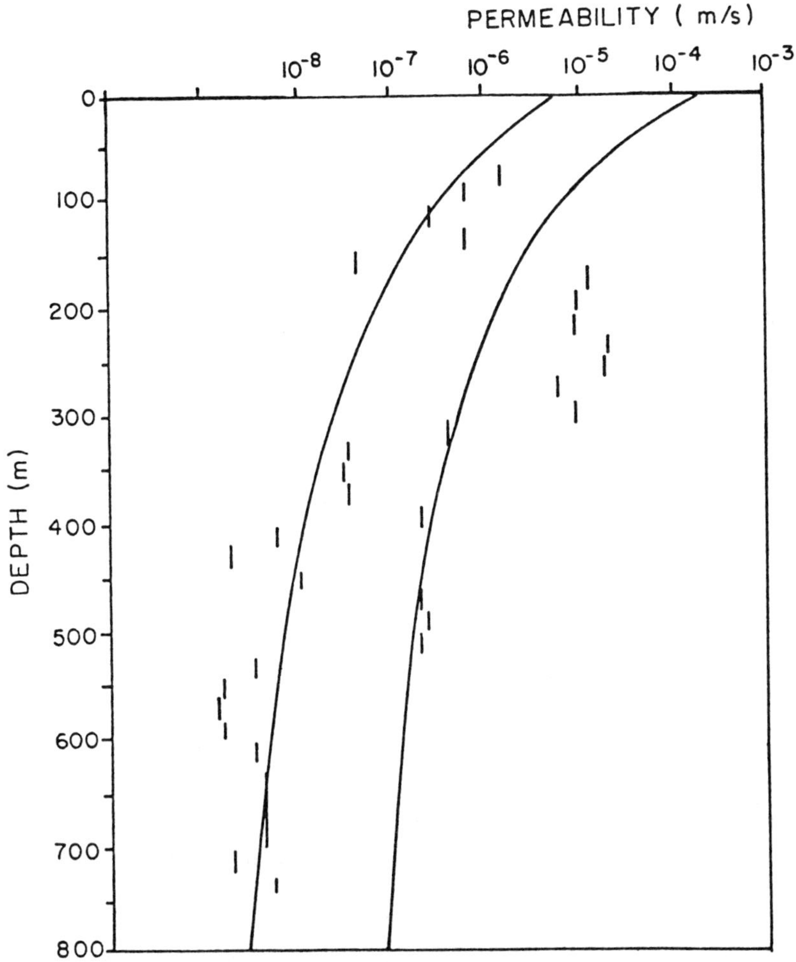

(b)

Figure 4.10 Measured and estimated conductivities for two boreholes (a) – Fi 5, (b) – Fi 6 in granodiorites (after Tiren *et al.*, 1985).

There are definite indications of a decrease in fracture porosity and fracture permeability of rock masses with depth at shallow depths. At these depths, which probably coincide with the active hydrological zone, models of hydromechanical coupling, normal closure and joint system geometry can be approximated for conducting aperture, porosity and permeability of *in situ* rock masses.

Connectivity, scale effects and percolation theory 5

It is difficult to analyse fluid flow through a discontinuous rock mass. Measured values of flow or permeability from field tests are affected both by the degree of interconnection of the discontinuity system and the heterogeneity of the individual discontinuities. Thus actual fluid flow may be more or less than that predicted by the simple analytical and empirical techniques described in previous chapters. To analyse fluid flow through a rock mass, a continuum model is commonly used. This assumes that the rock mass has a sufficient number of randomly orientated and interconnected discontinuities to simulate flow through a porous medium. Since a large number of macroscopic flow paths are needed to define, statistically, average flow properties, this approach is valid only if the average spacing frequency is much less than the size of the structure or the tested mass (Maini *et al.*, 1972; Wilson and Witherspoon, 1970).

Several recent studies (Hsieh and Neuman, 1985; Hsieh *et al.*, 1985; Neuman, 1987; Rasmussen, *et al.*, 1985; Andersson and Dverstorp, 1987) suggest that there may be limitations in conventional continuum models because of the inability to correctly account for the effects of natural discontinuity geometry. The evidence also indicates that the connectivity of a rock mass is affected by the discontinuity geometry (Oda, 1985; Long and Witherspoon, 1985; Andersson and Dverstorp, 1987).

Generally rock masses with high discontinuity densities and large discontinuity lengths will be well connected while rock masses with low densities and short discontinuity lengths will be poorly connected. The distinguishing threshold or range between good and poor connectivity is important in hydrological engineering, and if it can be determined from the discontinuity geometry, it may provide a sound basis for understanding the movement of fluid flow through rock masses.

A meaningful estimation of effective rock mass volume is also important. Many authors (Neuman, 1987; De Marsily, 1985; Long and Witherspoon, 1985) have discussed the validity of the Representative Elementary Volume (REV) concept in discontinuous rocks. REV can be used to identify an experimental scale which has the same characteristics as the fractured medium and to model a continuous equivalent medium.

In this chapter, the connectivity and scale effect of the rock mass are examined using the concept of **percolation theory** with a **percolation threshold**.

5.1 HETEROGENEITY AND INTERCONNECTIVITY

In a rock mass, discontinuities, whether they are joints, bedding planes or foliations, combine to form a discontinuous mass. When fluid passes through such a medium, flow becomes heterogeneous. Since the discontinuities act as channels, flow through the rock mass is controlled by the location, orientation and characteristics of individual discontinuities or discontinuity sets. In order to study fluid flow through a rock mass with impermeable or permeable intact rock between discontinuities, it is necessary to understand the nature of the heterogeneity and interconnectivity of the rock mass.

Long (1985) has proposed two criteria which must be satisfied if a heterogeneous system is to be replaced by an equivalent homogeneous system in a given dimension. First, there must be an insignificant change in the value of equivalent permeability with small additions to, or subtractions from, the flow volume. Second, a single equivalent symmetrical permeability tensor must exist which can predict the correct flux when the direction of the gradient in the REV is changed. However, there is evidence that the actual gradient or flux within a heterogeneous sample does not always have the exact symmetrical value of the flux in a continuum (Long and Witherspoon, 1985). Thus, the second criterion is not always satisfied since the rock mass is not always isotropic.

The degree of heterogeneity has been defined by Maini *et al.* (1972) as the ratio of the average spacing between discontinuities to the size of the structure under investigation. They suggest that if this ratio is greater than approximately 1/20, there will be serious errors in the value of the permeability coefficient if a continuum approach is considered. Wilson and Witherspoon (1970) suggest that the value should be less than 1/50.

Rouleau and Gale (1985) show that interconnectivity is a function of distributed discontinuity parameters with the orientation, size, spacing and density of discontinuity sets, used to define the connected proportion of discontinuities in the rock mass. Long and Witherspoon (1985) have shown that the interconnectivity of a rock mass affects both the magnitude and nature of discontinuity permeability. Interconnectivity between given sets is a complex function of **discontinuity density**, defined as the number of discontinuities per unit area or unit volume, and **discontinuity size**, defined in terms of area or length, which are not easily obtained from field investigations.

To quantify the degree of interconnectivity between two discontinuity sets, Rouleau and Gale (1985, 1987) proposed an empirical interconnectivity index, I_{ij}, illustrated in Figure 5.1 and given by:

$$I_{ij} = \frac{l_i}{s_i} \sin \gamma_{ij} \quad (i \neq j) \tag{5.1}$$

where l_i is the mean trace length for set i, s_i is the mean spacing of set i and γ_{ij} is

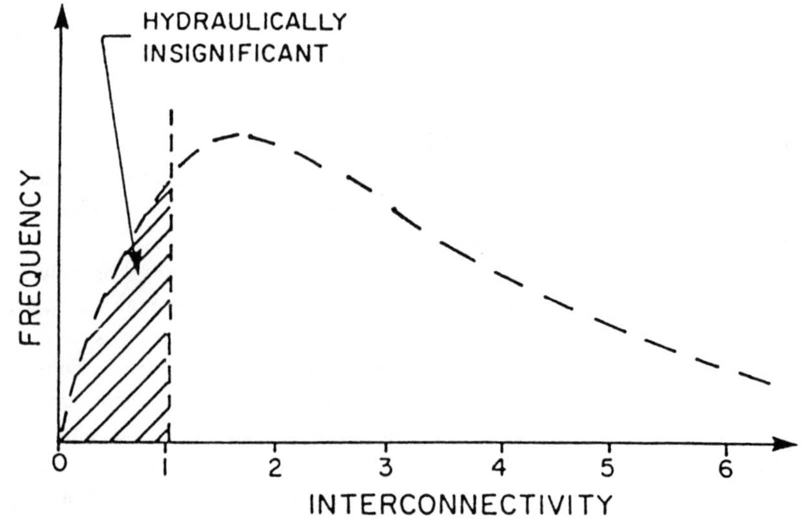

Figure 5.1 A hypothetical interconnectivity function for a discontinuity set (after Rouleau and Gale, 1985).

the average angle between fractures of set i and of set j. I_{ij} is generally different from I_{ji}. Rouleau and Gale (1985) also defined the total interconnectivity index for a set with an independent aperture as:

$$I_i = \sum_{j=1}^{n} I_{ij} \; (1 \neq j) \tag{5.2}$$

where n is the number of discontinuity sets. They show that a larger I_i has a greater hydraulic importance for set i. When an interconnectivity value is lower than a given threshold, these discontinuities do not provide hydraulic communication within the discontinuity network. The threshold value is equal to unity; i.e. when I_i is larger than unity, the ith set is connected in the discontinuity system.

5.2 PERCOLATION THEORY

Percolation theory requires a specific vocabulary. A **medium** is defined as an infinite set of abstract objects which are called **sites**. Fluid flows from a source site along paths connecting various sites. These paths are called **bonds**. These bonds may be orientated or non-orientated. An orientated bond permits the fluid to flow only in a specific direction while a non-orientated bond cannot allow fluid flow. The **co-ordination number**, Z, is the number of bonds leaving each site for a regular medium or periodic lattice. This number Z is dependent on the **nearest neighbours** or **next nearest neighbours** (Stauffer, 1985).

A random mechanism is generally selected to describe the percolation process.

This can be done in two different ways (Shante and Kirkpatrick, 1971; Stauffer, 1985). First, the bond percolation method describes how the random characteristics of the medium are selected by some of its bonds at random. Any bond, independent of all other bonds, has a fixed probability P_b of being closed or blocked. Second, the site percolation method assumes that all the bonds are open and may be orientated. The stochastic mechanism blocks sites chosen at random so that any site has a fixed probability (P_s) of being blocked.

For fluid flow problems, the bond percolation method is usually used. In this, fluid cannot flow through a blocked bond and it can neither wet nor pass through a blocked site. If a set of unblocked sites is connected one to another through unblocked bonds, a **cluster** will exist, which may be either finite or infinite. A finite cluster will only allow fluid to wet a finite number of sites. An infinite cluster will exist when the medium reaches a critical percolation condition. In this condition, there is a probability, $P(p)$, that fluid will wet or pass an infinite number of sites when $P(p)$ is selected at random and from a single source site. Mathematically

$$P(p) = \lim_{n \to \infty} P_n(p) \tag{5.3}$$

where $P_n(p)$ is the probability that a single source site will wet a least n other sites. To distinguish between the two probabilities $P(p)$ and p; $P(p)$ is the probability of an arbitrary site belonging to the infinite network, and p is the probability of an arbitrary site being occupied. The probability $P(p)$ and the probability p have been named as the **strength** of the infinite network and the **concentration** respectively (Stauffer, 1985). Here strength means the relative number of sites.

A critical probability (p_c) can be defined as:

$$p_c = \text{SUP}\ \{p, \text{ such that } P(p) = 0\} \tag{5.4}$$

Thus for $p < p_c$ the fluid may spread only locally, i.e. all clusters of wet sites are finite in size. And for $p < p_c$ at least one cluster of infinite size or a percolation channel can spread through the medium. Shante and Kirkpatrick (1971) have shown that for any given medium:

$$P_{ns}(p) < P_{nb}(p) \tag{5.5}$$

where subscripts s and b denote site and bond properties, respectively. If $P(p)$ is a monotonic increasing function of p, then $p_{cs} > p_{cb}$.

When p approaches p_c but is less than p_c the characteristic quantities of the cluster scales in regular lattices are expressed as:

$$L \sim |p - p_c|^{-v} \tag{5.6}$$

where L denotes the correlation length with corresponding critical exponent v. When p is just greater than p_c the conductivity of the network is assumed to behave like:

$$K_\sigma \sim |p - p_c|^t \tag{5.7}$$

where K_σ is percolation conductivity with corresponding critical index t. v and t are also called universal exponents for length and conductivity respectively in percolation literature.

The average critical length and critical conductivity exponents for a system of overlapping disks have been calculated in two dimensions. Values of v and t are as shown.

v	t
1.33 ± 0.07 (Vicsek and Kertesz, 1981)	1.13 ± 0.09 (Stinchcombe and Watson, 1976
132(+0.02, −0.07) (Cox and Essam, 1976)	1.38 ± 0.13 (Watson and Leath, 1974)
1.354 ± 0.015 (Reynolds *et al.*, 1978)	1.1 ± 0.1 (Kirkpatrick, 1973)
1.33 (Den Nijs, 1979)	1.0 (Wilke *et al.*, 1975)
1.33 ± 0.01 (Eschbach *et al.*, 1980)	0.9–1.5 (Englman *et al.*, 1983)
1.334 ± 0.002 (Nightingale and Bolde, 1980)	
1.28 for triangular lattice (Young and Stinchcombe, 1975)	
1.31 for square lattice (Young and Stinchcombe, 1975)	
1.33 (Wilke *et al.*, 1985)	

Wilke *et al.* (1985) suggest values of $v = 0.89$ and $t = 1.75$ for three dimensions and Englman *et al.* (1983) values of t between 1.6 and 2.0 for three dimensions in finite size specimens.

Gueguen and Dienes (1989) proposed a pipe model and crack model to account for permeability of low-porosity rocks with pipe or crack centres assumed to be distributed on a random network. By using a classical approximation, the critical fraction of connected pipes or cracks, f, is given by:

$$f \cong 54(p - p_c)^2 \tag{5.8}$$

where each pipe or crack has $Z = 4$ neighbours and p_c can be approximated by:

$$p_c \cong 1/(Z - 1) = 1/3 \tag{5.9}$$

Stauffer (1985) has indicated that critical percolation generally occurs where concentration probability, p, at an infinite network appears in an infinite lattice. This probability is based on investigations of regular lattices for the case of percolation in random structures. However, in finite systems as simulated on a computer, there is rarely any sharply defined critical value. Any effective critical values obtained numerically or experimentally need to be extrapolated to the infinite system. If a mathematically exact calculation for p_c exists, then no such extrapolation is needed. Mathematical methods to calculate this critical value exactly are often restricted to two dimensions and regular lattices. This is consistent with experience in the field that three dimensional problems cannot be solved exactly.

5.3 PERCOLATION THRESHOLD AND INVARIANTS

The determination of a critical concentration probability p_c, for the occupation probability, p, $(0 \leqslant p \leqslant 1)$, is important for macroscopic random media studies. It has been shown from computed values of p_c obtained for site and bond percolation on periodic lattices that there exist dimensionally invariant combinations of p_c with the number co-ordinance, Z, of the lattice, or with the filling factor, f_s, for densely packed discs in two dimensions and for spheres in three dimensions (Shante and Kirkpatrick, 1971; Charlaix *et al.*, 1984; Stauffer, 1985). These invariants and periodic lattices, illustrated in Figure 5.2, have been observed to depend only on the dimension of the lattice and not on the specific lattice types. In Figure 5.2 it can be seen that the most widely used invariants are $B_c = Zp_{cb}$ with $B_c = 2$ in two dimensions and 1.5 in three dimensions for the bond percolation method, and $V_c = f_s p_{cs}$ with $V_c = 0.45$ in two dimensions and 0.15 in three dimensions for the site percolation method. The invariants B_c and V_c are represented as the critical values for the average number of bonds per site in bond percolation and the fractional volume associated with active discs or spheres in site percolation (Charlaix *et al.*, 1984). Other studies, discussed below, have dealt with fracture density and fracture length as well as with threshold values.

Watanabe (1986) showed that the threshold value of any type of infinite network can be approximated by the form:

$$Zp_c \cong D/(D-1) \tag{5.10}$$

where Z is number of bonds initiating from each site, D is the dimension of the network (e.g. $D = 2$ for a two-dimensional network). This equation shows that the critical probability depends on the network geometry, but that the product of Z and p_c does not. However, when the number of dimensions is more than three, this formula is questionable (Madden, 1983). Equation 5.10 implies that the invariant is 2 and 1.5 in two and three dimensions, respectively.

Vicsek and Kertesz (1981) propose an irregular lattice made up of discs of equal radius, r. The centres of these discs (sites of the random lattice) are uniformly distributed on a plane with density, ρ. Then a dimensionless variable, p, called the area fraction, which plays the role of occupation probability, can be expressed by:

$$p = 1 - \exp(-\pi r^2 \rho) \tag{5.11}$$

One disc can be partially or fully overlapped by at least one other disc. Vicsek and Kertesz (1981) found the critical area fraction $p_c = 0.668 \pm 0.005$.

Robinson (1982) assumed that the orientation of fractures in two dimensions was purely random and found that the definition of the percolation threshold could be related to fracture density and fracture length:

$$[N < r^2 >]_c \cong 1.54 \tag{5.12}$$

where N is the fracture density (number of fractures per unit area), r is the length of fractures, c denotes the critical value, and $< >$ denotes the average value.

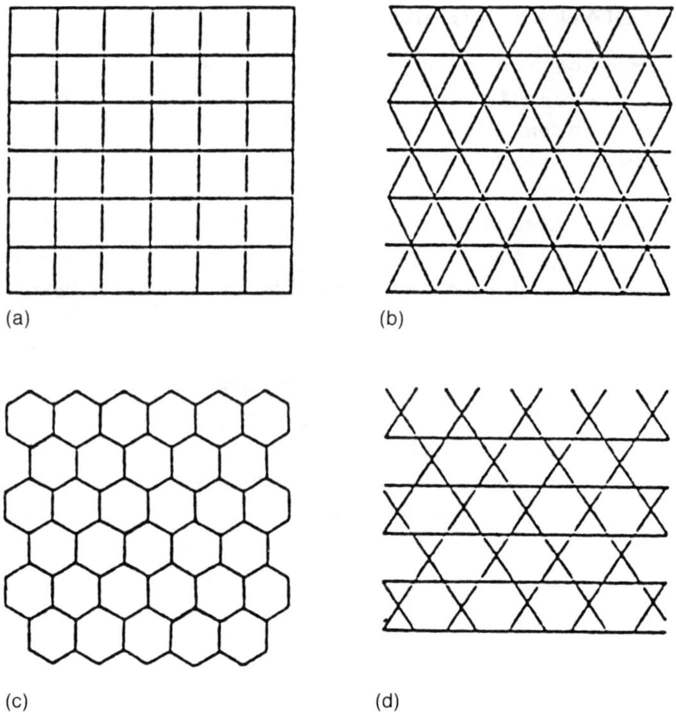

Figure 5.2 Some common two-dimensional lattices and related critical probabilities and dimensional invariants (after Kirkpatrick, 1973): (a) square; (b) triangular; (c) honeycomb; (d) Kagomé.

Lattice	Z	p_{cb}	Zp_{cb}	f_s	p_{cs}	$f_s p_{cs}$
Honeycomb	3	0.6527	1.96	0.61	0.700	0.427
Kagomé	4	—	—	0.68	0.653	0.444
Square	4	0.5000	2.00	0.79	0.590	0.466
Triangular	6	0.3473	2.08	0.91	0.500	0.455
Diamond	4	0.388	1.55	0.34	0.425	0.145
s.c.	6	0.247	1.48	0.52	0.307	0.160
b.c.c.	8	0.178	1.42	0.68	0.243	0.165
f.c.c.	12	0.119	1.43	0.74	0.195	0.144
h.c.p.	12	0.124	1.49	0.74	0.204	0.151

Charlaix *et al.* (1984) proposed a percolation threshold for a random array of flat discs in three dimensions, which can be expressed as:

$$[\tfrac{1}{2}N < \text{surface area of fracture} > < \text{perimeter} >] \tag{5.13}$$

where N is the fracture density (number of fractures per unit volume) and its value is in the range from 1.5 to 3.0. When the fracture is looked upon as a circular disc of radius, r, it becomes:

$$[N\pi^2 < r^2 > < r >]_c \cong 1.5 \text{ to } 3.0$$

i.e.

$$[N < r^2 > < r >] \cong 0.15 \text{ to } 0.3 \tag{5.14}$$

Meanwhile, when all fractures have the same radius r, it becomes:

$$[N < r^3 >]_c \cong 0.15 \text{ to } 0.3 \tag{5.15}$$

Wilke *et al.* (1985) and Charlaix *et al.* (1984) carried out numerical calculations using unit squares on a percolation problem with a cubic lattice. They obtained a range of percolation threshold values:

$$[Nr^3]_c \cong 0.24 \text{ to } 0.09, \text{ i.e. } [N\pi^2 r^3]_c \cong 2.4 \text{ to } 0.9 \tag{5.16}$$

Englman *et al.* (1983) reported that the percolation frequency, υ, the ratio of percolation networks to the total number of fracture networks, was a function of sample size, R, and effective bond-occupation probability, ξ, also called the percolation factor. This relation is given by:

$$\upsilon = f(R, \xi) \tag{5.17}$$

where $\xi = \pi N_A l_s^2$, l_s is the length of fractures or cracks and N_A is the number of fractures per unit area. The relation of percolation factor to percolation threshold is discussed in the next section.

Long and Witherspoon (1985) have discussed a length – density (LD) fracture method. They proposed a probabilistic relationship for fracture frequency υ expressed by:

$$N_L = N_A l_s \cos\theta$$

$$\frac{N_L}{\cos\theta} = N_A l_s = LD \tag{5.18}$$

where N_L is fracture frequency, N_A is the density (number of fractures per unit area or volume), l_s is the length or area of the fracture and θ is the angle between the sample length and fracture poles of the given set.

This relation indicates that it is not necessary to look at all possible combinations of fracture length or area and density directly. The product $N_A l_s$ may be estimated since N_L and θ can be estimated relatively easily from direct borehole measurements. If the product, $N_A l_s$, is constant, then the threshold can be measured using Monte Carlo simulations. Long and Witherspoon also show that the relation between measured scale and permeability is related to percolation threshold.

5.4 PERCOLATION FRACTURE[1] FLOW

The basic concept of fluid flow in hard discontinuous rocks – that the rock matrix is impermeable – has been discussed by many authors (Long and Witherspoon, 1985;

[1] In discussion of percolation theory the term 'fracture' is often more appropriate than 'discontinuity' and will be used in most sections.

Charlaix *et al.*, 1984; Englman *et al.*, 1983; Wilke *et al.*, 1985; Watanabe, 1986). In percolation theory, the individual discontinuities are conceptually modified to become finite extent fractures of variable individual permeability related to the aperture of the fracture. The fractures may then form a percolation network or a bond percolation model of intersecting fractures. The original network, for instance, in two dimensions, defined by the number of bonds, Z, initiating from each site, is composed of many sites linked by bonds. Then the modified network is constructed by moving out some of the bonds from the original network in order to decide whether each remaining or removed bond is independently established using a fixed existence probability which is a function of a factor dependent on the size and/or density of fractures in the medium. This factor, ξ, is called the **percolation factor**. When the size or density of fractures is seen as a continuously increasing step in the medium, the density or size of fractures between which enough intersections exist for general fracture flow to be possible is reached at a critical value of percolation factor, ξ^*, called the **percolation threshold**. This value is generally regarded as the distinguishing range of connection or non-connection in the medium.

Below the threshold, fractures will not be interconnected, and thus no flow will be possible in the medium. At this stage ($0 < \xi < \xi^*$), a finite set of fractures can be interconnected, creating a **finite cluster**. Although no general fracture flow can occur in the medium, each cluster can be considered locally permeable. When the size or density is increased again, the percolation threshold is reached. At this stage, at least one cluster will become infinite. In other words, at least one cluster is completely connected throughout the medium and becomes a pathway through which water can flow. Above the percolation threshold, the interconnectivity of the medium increases with increasing size or density of fractures. Local finite clusters may still exist independently of the infinite clusters.

Due to the complex nature of discontinuities within a rock mass, an exact mathematical treatment of the relation between interconnectivity and percolation factor for the percolation network with a randomly located factor is not available. Thus it is difficult to obtain an exact calculation of p_c and p, especially in three dimensional rock mass problems. The effect of scale is also important in analyses of fracture flow using percolation theory. Equation 5.17 illustrates the relations between conductivity, percolation threshold and scale effect, and the comments above can be expanded to explain this:

1. When ξ is below the percolation threshold, ξ^*, increasing the scale of measurement will decrease the percolation frequency, υ. Thus, an equivalent continuous medium cannot be defined. Local finite clusters of fractures may be conduits and may be connected, but two separated clusters will only be linked either by the permeability of the rock matrix or by the minor sets of connected fractures. De Marsily (1985) states that at this stage, the conductivity of a rock mass may be determined in two ways: the first by exhaustive sampling and description of all the fractures in space; the second by using Monte Carlo simulation to generate a large number of networks in order to statistically determine the probability of flow taking place in a finite cluster between two points of interest.

2. When $\xi > \xi^*$, increasing the scale of measurement increases the percolation frequency. It is possible to select a scale at which the medium can be an equivalent continuous medium and the size of REV may be defined.
3. An infinite sample such as the *in situ* case with $\xi > \xi^*$ has a percolation frequency equal to unity. Thus an *in situ* hydraulic test on a finite sample can be used to characterize global behaviour.
4. An infinite sample with $\xi < \xi^*$ has a percolation frequency equal to zero. Thus an *in situ* hydraulic test on a finite sample cannot be used to characterize global behaviour.
5. As ξ approaches ξ^*, the percolation frequency will become either unity or zero (either permeable or impermeable). De Marsily (1985) states that in this situation, the hydraulic properties of the network can be characterized by general relations between universal constants (i.e. independent of the type of network). In other words, if the sampling shows a spatial variation of fracture density and length, its influence on the probability of connection between two points, or on global conductivity, can be predicted.

It is therefore necessary in any analysis to first define the boundaries of the fractured rock medium. Then key equations including distribution of fracture density, orientation and length, can be used to define the percolation threshold and to compare the actual fracture density or size to this threshold. For scale problems concerning flow through a rock mass, Sagar and Runchal (1982) have concluded that 'any fracture which does not appear on the boundary of a rock element considered is of no interest in the calculation of the equivalent permeability'. This implies that if a finite boundary or area cannot be defined, it is not possible to determine the equivalent behaviour of a rock mass which is strongly controlled by geological features.

5.5 ESTIMATION OF ROCK MASS PERMEABILITY BY PERCOLATION THEORY

An example of the use of percolation theory to estimate permeability is the disc-type fracture model with disc fractures (Figure 4.5) developed to obtain the permeability of a rock using statistical methods by Dienes (1982) and Gueguen and Dienes (1989). This model shows that permeability and porosity are functions of an isotropic distribution of fractures with average radius, r, average half-crack aperture, a, and average crack spacing, s, given by:

$$n = 2\pi \frac{r^2 a}{s^3}$$

$$k = \frac{4\pi}{15} \upsilon \frac{a^3 r^2}{s^3} \tag{5.19}$$

where the factor υ is the percolation frequency that belongs to an infinite network (equation 5.17) and is determined from percolation theory; $1/s^3$ is the crack density (number/volume). Gueguen and Dienes (1989) propose a classical approximation using a Bethe lattice with four co-ordinates ($Z = 4$) to calculate the percolation frequency, υ. They show that if υ is near the percolation threshold, then the critical percolation frequency is given by:

$$\upsilon \cong -54(p - p_c)^2 \tag{5.20}$$

where $p_c = 1/3$. Since υ is known to be a function of a probability, p, in the percolation literature, it can be calculated exactly from p with simple lattices (section 5.3). Gueguen and Dienes (1989) assign a probability in terms of two intersected fractures. The probability p for two fractures to intersect is determined by:

$$p = \frac{\pi^2 r^3}{4s^3} \tag{5.21}$$

Equation 5.19 also indicates that $K = K_{max}$ is given when $\upsilon = 1$, i.e. when the discontinuities in the medium are well connected. Equations 5.20 and 5.21 indicate that flow will only occur when $p > 1/3$ and $r/s > 0.5$.

An example of permeability and porosity estimation using this approach can be shown by use of data from granite at Stripa, Sweden (Gale *et al.*, 1982; Wilson *et al.*, 1983) to compute permeability and porosity. The computed permeability and porosity are then compared with measured data. Stripa mine was the site of a major US–Swedish experiment to determine the feasibility of geologic repository construction in crystalline rock, and is the source of considerable structural data. The discontinuity orientation data showing four sets of joints, from Rouleau and Gale (1985), are summarized in Figure 5.3. Table 5.1 summarizes the mean orientation, mean trace length, mean spacing and mean trace radius of the four sets of discontinuities found at the Stripa site. The mean discontinuity radius of each set in the table is calculated using Robertson's (1970) equation

$$r_i = \frac{4}{\pi} L_i \tag{5.22}$$

where L_i is the expected value of the trace length and r_i is the expected value of the trace radius.

The average trace length, average discontinuity spacing and average trace radius are 1.38 m, 0.65 m and 0.88 m, respectively. These values were computed by taking the average of the four sets.

The geometric mean of the intrinsic permeability of the rock mass at the Stripa site is 8.9×10^{-5} darcies. This mean value was calculated from 148 individual tests in two boreholes by Gale *et al.* (1982). The porosities are reported as typically less than 10^{-4}. The mean discontinuity aperture is 8.3 μm, which is the average value from 2 m interval tests (Gale *et al.*, 1982). The rock mass permeability and porosity at the Stripa site were calculated from the average input parameters from Table 5.1 using the three models in Figure 4.5. The percolation frequency, υ, of the disc-type fracture model is assumed to be 1.0 since the percolation factor of the Stripa site is significantly greater than the percolation threshold of 2.4 discussed in Chapter 6. The percolation

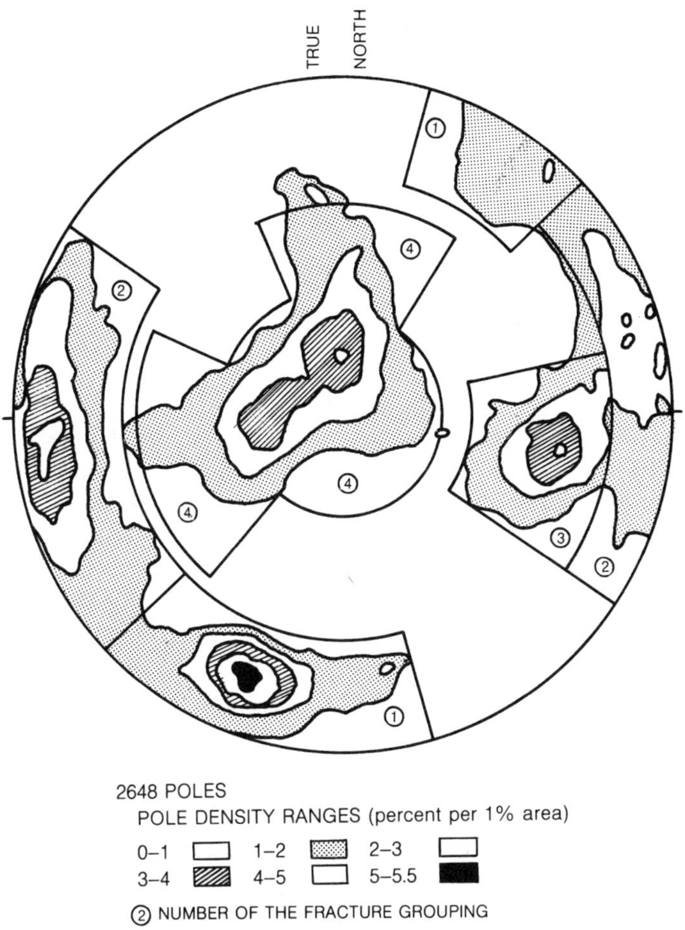

Figure 5.3 Discontinuity orientation data plotted as poles on a lower hemisphere projection (after Rouleau and Gale, 1985).

Table 5.1 Mean value of fracture parameters at the Stripa site (data from Rouleau and Gale, 1985)

	Set 1	Set 2	Set 3	Set 4	Average
Mean trace length (m) (exponential model)	2.16	0.83	1.51	1.03	1.38
Mean spacing (m)	0.93	0.36	1.61	0.51	0.65
Mean trace radius (m)	1.38	0.53	0.96	0.66	0.88
Mean dip direction	23	83	278	155	–
Mean dip	76	85^	53	12	–

1. ^ denotes that original value was 95 (Gale and Rouleau, 1982).
2. Average percolation factor = 24.3 > 2.4.

Table 5.2 Comparison of fracture permeability and fracture porosity calculated from joint system models and obtained from field measurements by Gale *et al.* (1982)

	Fracture permeability (darcy)	Fracture porosity
One parallel set	0.73×10^{-4}	1.3×10^{-5}
Three orthogonal sets	1.47×10^{-4}	3.8×10^{-5}
Disc-type fractures	1.68×10^{-4}	7.4×10^{-5}
Field measurement	0.89×10^{-4}	$< 10^{-4}$

factor of the site was calculated from:

$$\xi = \pi^2 r^3/s^3 = \pi^2(0.88/0.65)^3 = 24.3 \tag{5.23}$$

Calculated values from equations 5.19–5.21 and field measured permeability and porosity values are compared in Table 5.2. This shows that the permeability calculated from models with three orthogonal sets and disc-type fractures are slightly higher than the field measured values. However, the values are of the same order of magnitude. The calculated values using the three orthogonal sets and disc-type models are in close agreement. All calculated values of porosity using the three models are less than the 10^{-4} reported by Gale *et al.* (1982).

5.6 REPRESENTATIVE ELEMENTARY VOLUME (REV)

It has been shown earlier that in order to obtain a value of rock mass conductivity representative of the rock mass, tests must be carried out on a sample size which is equally representative. If tests were carried out on samples of variable volume, it would be expected that as the sample volume increased, the measured conductivity would eventually tend to a constant value. At this volume the sample volume becomes representative of the rock mass, and it is called the Representative Elementary Volume or REV (Figure 5.4(a)). When the sample volume is less than REV, the unrepresentative fraction of discontinuities sampled will determine the measured conductivity. When the volume is greater than or equal to REV, fluctuations in the measured conductivity will be less, and the values will be representative (Figure 5.4(b)). Unfortunately, the complex nature of rock discontinuity systems means that REV is difficult to define.

Many authors have discussed the concept of REV (Long *et al.*, 1982; Long and Witherspoon, 1985; Neuman, 1987; Kunkel *et al.*, 1988). Long *et al.* (1982) show that REV may exist on several scales. For instance, REV for the equivalent hydraulic conductivity tensor is related to the boundary conditions and the gradient under which flow occurs. They concluded that the size of REV depends on the flow system and the geometrical parameters of the discontinuous medium.

Kunkel *et al.* (1988) showed that REV will increase in size with discontinuity spacing. Figure 5.5 illustrates how fracture sets and hydraulic features are related to volume. In unfractured rock, REV could be any reasonable size because the hydraulic properties of the system do not change significantly if a small volume of rock mass

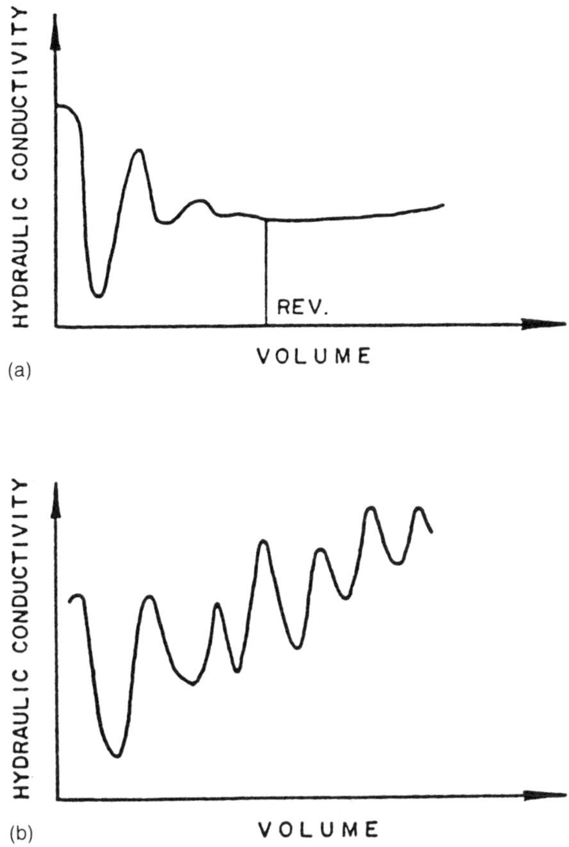

Figure 5.4 Statistical representation of homogeneity and REV: (a) in unfractured rock; and (b) in fractured rock (after Kunkel *et al.*, 1988).

is added to or subtracted from REV. In discontinuous rock, REV should be large enough to include sufficient fracture intersections to maintain constant values of hydraulic properties if a small volume of the rock mass is added to or substracted from REV. For large-scale features, REV may not exist or may be so large, for instance several cubic miles, as to be irrelevant. This implies that the hydraulic properties continue to change with size, and that REV may or may not exist for every permeable system, depending on the frequency of discontinuity intersections and geometrical parameters.

Thus, the REV concept may not be satisfied for every rock mass. The concept depends on the identification of complex clusters of discontinuities in a rock mass (Wilke *et al.*, 1985; De Marsily, 1985; Neuman, 1987). The only way to define REV for a rock mass is to investigate in detail the discontinuity geometry. There is also a difficulty when flow is controlled by larger discontinuities or faults. Thus, if

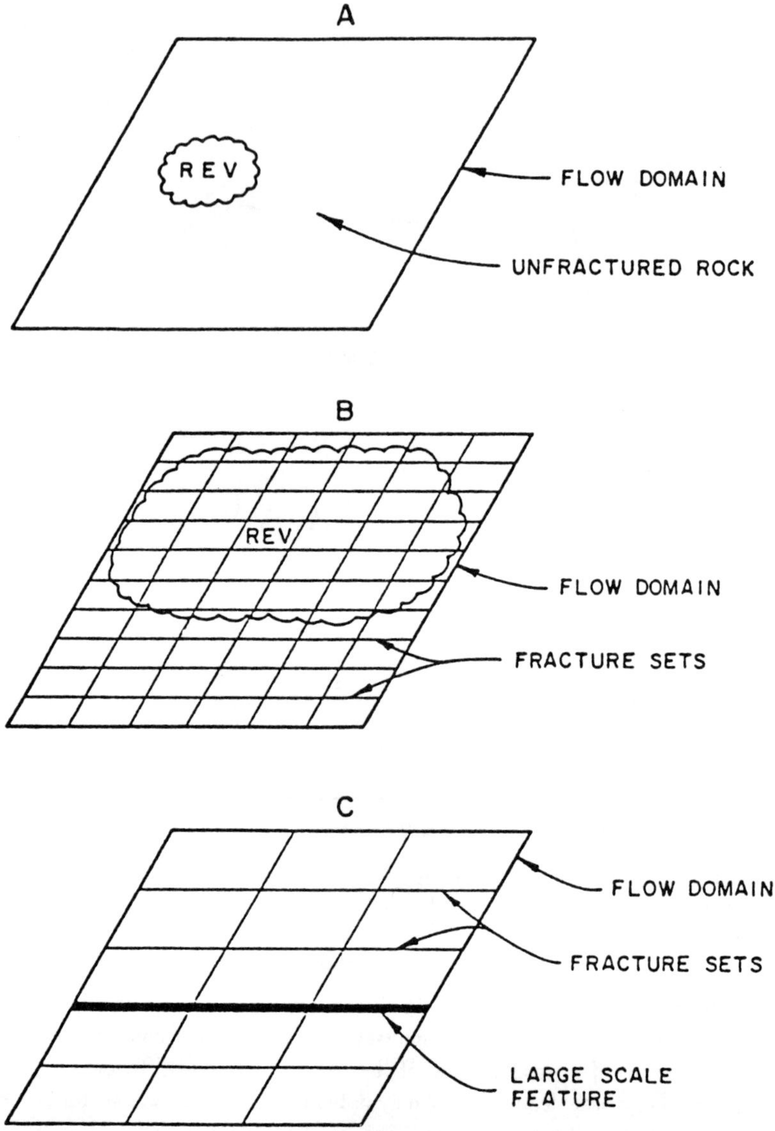

Figure 5.5 Conceptual discrete hydraulic features related to REV: (a) in unfractured rock; (b) in fractured rock where REV includes sufficient fracture intersections to represent the mass flow domain; and (c) in fractured rock where large-scale features mean that REV is very large or non-existent (after Kunkel *et al.*, 1988).

experimental results show that rock mass porosity or permeability exhibit a kind of star-shaped diagram when using continuum analysis, this invalidates the REV concept (Panel Discussion, 1985). It is also evident that when REV is below the percolation threshold and a discontinuity system involves clusters, it is difficult to extrapolate from a small to a large region (De Marsily, 1985), and the REV concept cannot be justified. In this case extensive sampling is needed to describe the nature of the discontinuity system in space.

Flow models 6

One approach to improving data resolution is to develop a flow network model to which appropriate boundary conditions may be applied. Then established numerical techniques may be used to predict steady-state flow. The complexity of the problem often limits the application of analytical techniques to global domain geometry. However, if efficient simulation techniques can be found to provide sensitivity analysis, the effects of discontinuity parameters and stress on flow properties can be examined using a systematic basis for a discrete fracture system and using numerical simulation techniques.

6.1 FLUID FLOW MODELS

Numerical models using finite difference, finite element and boundary element methods have been used to model fluid flow through discontinuous rock masses. These models have been based on discrete fracture flow, dual porosity and continuum analogies of fluid flow.

Dual porosity models have been developed where substantial flow through both rock matrix and discontinuities occur simultaneously (Shapiro and Andersson, 1983; Rasmussen, 1988). The proportions of fracture flow and matrix flow can be determined by solution of two sets flow equations using coupling parameters to represent flow between the matrix and fractures. The head and flux should be balanced in these two flow domains.

Conventional **continuum** models are widely used for predicting fluid flow through low permeability rocks. It is assumed that fluid flow through a rock mass is similar to flow through a porous medium (Long *et al.*, 1982; Oda, 1986). Then an equivalent porous medium (EPM) formulation which assumes that a uniform porous material consists of a large number of macroscopic flow routes is designated by average macroscopic parameters. Although the EPM concept has been applied to discontinuous rocks (Neuzil and Tracy, 1981; Long and Witherspoon, 1985; Oda, 1986), EPM formulations are only valid if the number of discontinuities is sufficient to allow determination of statistical average values of flow paths. The scale in the statistical rock mass needs to be large, so that the medium can be treated as a homogeneous continuum (Neuman, 1987). The measurement scale must be chosen to obtain sufficient numbers and lengths of discontinuities and the scale of the test volume must reach REV in the continuum model.

In reality, there is evidence (section 5.6) that a rock mass containing less than a certain number of discontinuities may never reach the required REV. Flow-related changes in effective stress and fracture surface properties may further affect flow patterns, resulting in changes in fracture aperture related hydraulic behaviour (section 3.5). For instance, it is not easy to modify the consequences of shear-related dilational opening of discontinuities caused by changes in effective stress (section 3.2). Moreover, the degree of connectivity of fractures and superconnection with faults must not be overlooked in the continuum case. Thus, a continuum model involving conventional tensor analysis of an anisotropic rock mass may not be enough to evaluate flow within a rock mass. Although models of coupled stress and fluid flow have been treated as anisotropic elastic porous media with corresponding elastic compliance and permeability tensors, this is only valid in two-dimensional numerical analysis (Oda, 1986). Incorporation of these factors into any analysis of flow through fractures with an impermeable rock matrix, or even with a permeable rock matrix, is only possible in a discrete fracture model.

Discrete fracture flow models in two dimensions (Long and Witherspoon, 1985; Robinson, 1982) and in three dimensions (Huang and Evans, 1985; Rasmussen, 1988; Andersson and Dverstorp, 1987; Dverstorp and Andersson, 1989) have been developed. Long (1985) and others have extended a two-dimensional discrete fracture model to three dimensions by using analytical solutions for boundary conditions within circular fractures of uniform aperture. They provide a solution for complete three-dimensional flow which allows for validation with other simulation models. Similarly, Huang and Evans (1985) have proposed a method for simulating a wide range of field conditions by using a three-dimensional formulation of the fracture network with a one-dimensional flow tube formulation within the plane of the fractures. Their model was used to determine the hydraulic conductivity of the flow region between all intersections of each fracture. The pressure head in this case is applied at the exterior boundary of a simulated rock mass; all discontinuities are idealized to be smooth, parallel plates, and flow through them is assumed to be laminar and to obey Poiseuille's law. One disadvantage of the Huang and Evans model is that multiple sources and sinks within a discontinuity are not incorporated, and the influence of orientation is not incorporated. Rasmussen (1988) used this model together with the boundary integral element method to determine the relative contribution to total fluid flow through a rock block of flow through the rock matrix and through discontinuities. His model allows coupling of source and sink terms along the interface between the fracture wall and the rock matrix. However, in his model, the sources and sinks only exist within a fracture. And he does not consider the case where the fracture intersects the boundary and the flow occurs along the boundary.

From these studies it may be concluded that, whether a rock mass behaves like a continuum or a discrete fracture model, the discontinuity size and frequency are important factors. Thus, the estimation of flow behaviour in rock masses requires a detailed description of discontinuity geometry.

In general, discontinuities are grouped in sets on the basis of their orientation. Within a set, the variability in orientation, spacing, trace length and size can be observed. Since the field measurement of all discontinuities is difficult, a statistical approach is needed. Thus geometrical parameters can be analysed by using stochastic models. These require (Chapter 2) a rock mass fracture network that can be generated and used as input in a flow problem. From known statistical distributions of aperture, density, trace length and orientation, numerical techniques may be used to construct a representative rock mass as a series of Poisson discs.

Some authors have studied the effect of discontinuity parameters on the variation of hydraulic properties (Andersson and Dverstorp, 1987; Long and Witherspoon, 1985). Andersson and Dverstorp (1987) indicate that when the statistical variation of the flow rate is small, the results of using the discrete fracture model will be good. When the variation of flow rate is large, the results from the discrete fracture model are more variable. They suggest that a condition simulation model with known discontinuity geometries may predict flow better than a network of discrete fractures. They also show that conditional simulation decreases the prediction uncertainty of flow if fractures are long and fracture density is high. On the basis of the connectivity of fractures in two dimensions, Long and Witherspoon (1985) showed that the longer the discontinuities and the higher their density in a rock mass the closer its behaviour will approach that of a porous medium. This indicates that when a rock mass has longer discontinuities with higher density, the uncertainty of flow will decrease.

Appropriate boundary conditions must be applied to the assembled network. Since the complexity of the problem often limits the applicability of analytical techniques, numerical techniques are used to accurately represent global domain geometry. To consider the influence of all the factors mentioned above, the discrete fracture system may use conditional simulation techniques to examine the sensitivity analysis. Although numerical analysis may provide an invaluable and indispensable tool for validation, experimental and *in situ* characterization methods are difficult to justify on the basis of both their cost and the inherent difficulty of defining boundary conditions. This leaves simulation techniques as potentially the most attractive (Andersson and Dverstorp, 1987).

6.2 DISCRETE FRACTURE FLOW MODEL FOR CONNECTIVITY

In order to determine if a fracture network can be treated as connective or non-connective, a simulation model with a three-dimensional mesh generator can be used. This type of numerical program is designed to generate a fracture or discontinuity system and to measure the percolation factor and the percolation frequency of the system. Monte Carlo simulation can be used to illustrate the effects of fracture length and density on percolation factor and percolation frequency. The fracture system can thus be observed and the way in which its behaviour approximates a connective or non-connective medium can be demonstrated. The flow chart of a suitable program is described in Figure 6.1.

The three-dimensional fracture model simulates a fracture population. Input to the model includes specification of the possible distributions of fracture geometry parameters, which represent the statistical properties of a fracture population as

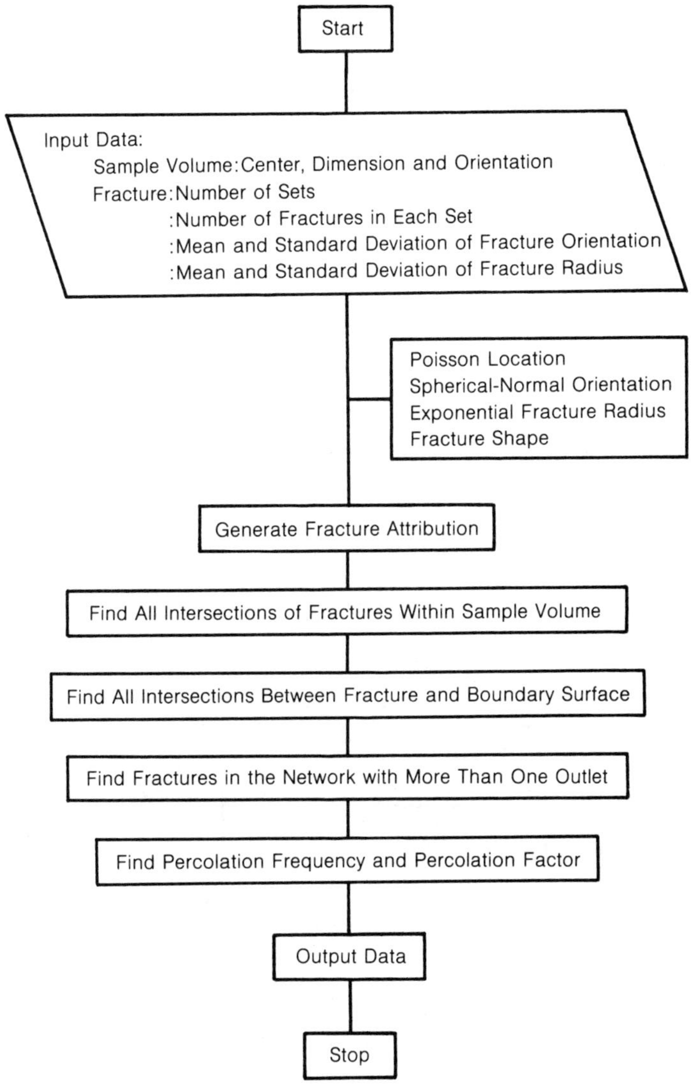

Figure 6.1 Flow chart for a discrete fracture model.

outlined in Chapter 2. The model can randomly generate fractures for a finite volume according to the specific statistical properties of a fracture population. Once the fracture pattern is generated in a global volume, a specific sample volume within the global volume is then subjected to numerical tests. This sample volume can be equal to the global volume. Thus, the interconnectivity of sample volume can be determined from the percolation factor and percolation frequency. This type of model is similar to the models of Long and Witherspoon (1985), Rasmussen (1988, 1991) and Dverstorp and Andersson (1989). It can provide a semi-analytical means for estimating the effective point or range of interconnectivity on a given scale. Additionally, the program can input the geometrical data of each fracture if these are known.

Mesh generation requires simplification of fracture geometry. A solid rectangular volume is treated as a matrix within which fractures are generated. Fractures are represented as a two-dimensional finite plane. The matrix is defined using a global volume and a sample volume with height, width and length (Figure 6.2). The global volume and sample volume are composed of six exterior and interior surfaces, respectively. The sample volume is extracted from inside the global volume. In selecting the sample volume, it is necessary to remove the effect of under-sampling near the boundary of the global volume.

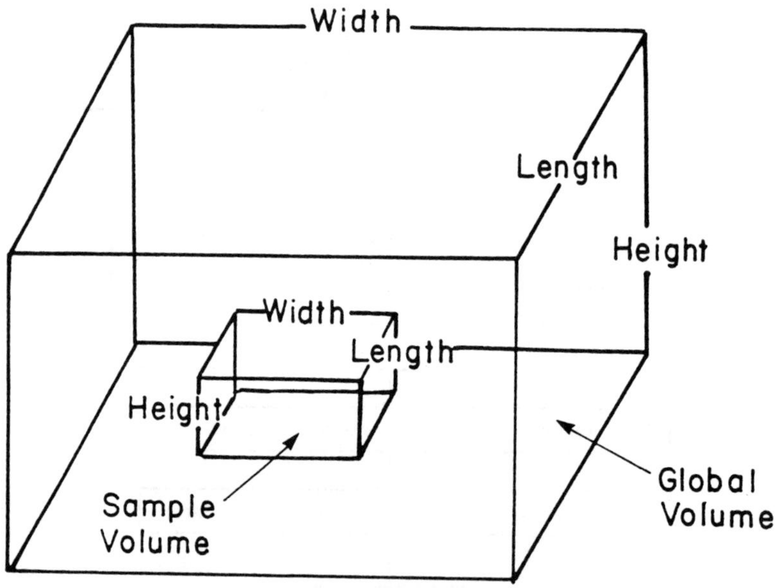

Figure 6.2 Definition of global and sample volume.

Once the global volume is determined, the mesh generating process can be used. The generated fracture network involves a number of individual finite planes within a three-dimensional global volume (Figure 6.3). Fracture patterns are produced according to the best currently available description of real fracture systems from geological field data in the area. The sets of fractures are assumed to be independent.

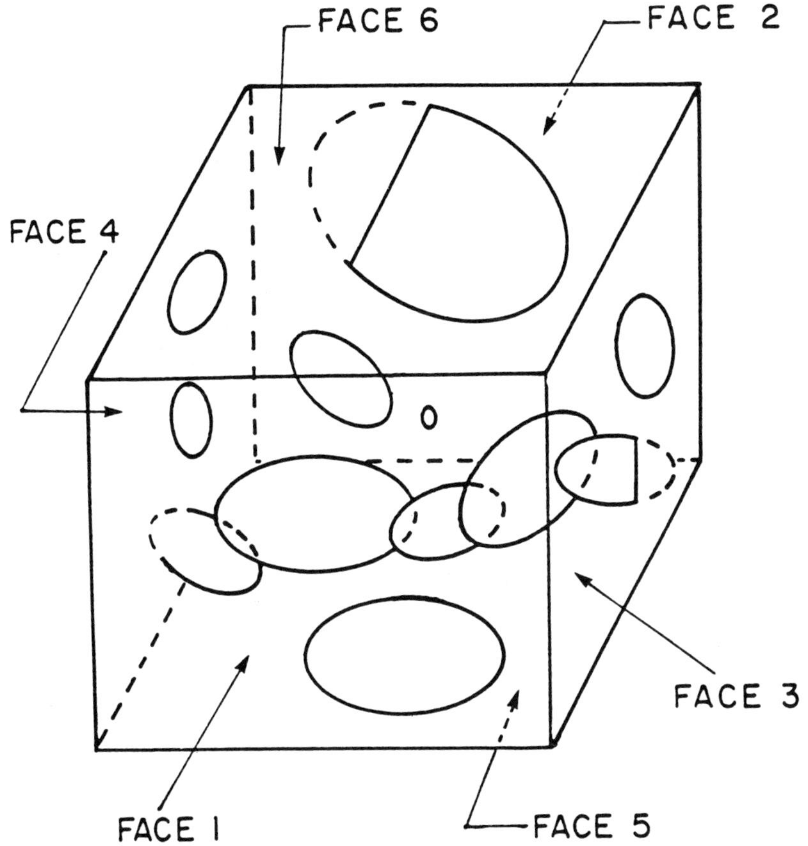

Figure 6.3 Three-dimensional fracture model with circular fractures.

The centres of individual fractures in global volume are randomly located by the Poisson process. When the fracture pattern is randomly generated in the global volume, the sample volume can be extracted from the global volume.

The location of each fracture in each set is generated by assuming that the centres of the fractures are random according to a Poisson process. These centres must be within the global volume and be obtained by a uniform probability taking pairs of pseudo-random numbers between 0 and 1. The co-ordinates of the fracture centre points can be obtained when the probability is multiplied by the height, length and width of the global dimension. During this process, the distance between fracture centres will result in an exponential distribution.

The orientation of each fracture in a set is assumed to be a spheric normal. The mean and variance of the orientation must be supplied for each set (section 2.2). The orientation of each fracture plane (Figure 6.4) is defined by a horizontal angle, B_p, and vertical angle A_p. B_p is measured anti-clockwise from the $+X$ axis. A_p is measured

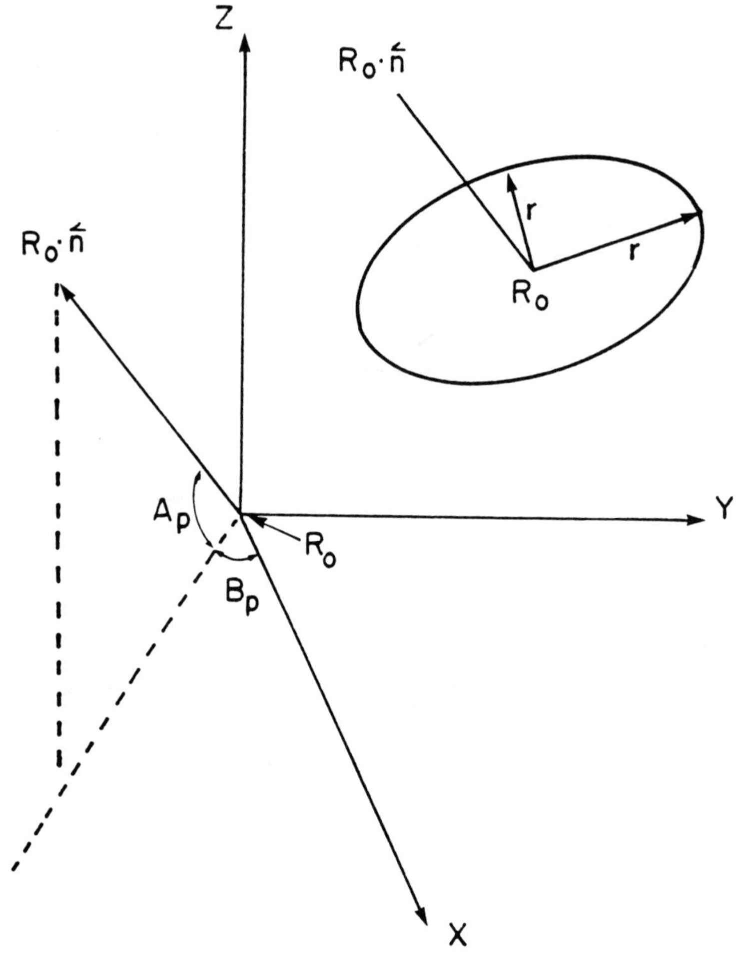

Figure 6.4 Position and orientation parameters for discrete fractures (modified after Rasmussen, 1988).

from the X–Y plane. These can be transferred from the strike and dip of a fracture set (section 2.2). Once the orientation is determined, then the infinite plane equation involving the finite fracture plane can be determined by these two angles (Rasmussen, 1988).

The fractures are assumed to be circular. The fracture length within each set is assumed to be exponentially distributed. Thus, the fracture radius gives an exponential distribution (section 2.3), and the expected value of fracture radius, L, must be supplied. During the generating process, part of the fracture radius may be outside the global volume. Thus, parts of these fractures are truncated at the boundary of the global volume, so that parts of their areas and perimeters also fall outside the global volume.

In this case, estimation of fracture area and perimeter must involve cutting off the parts outside the global volume.

Once all sets have been generated, intersections between fractures and between fractures and surfaces of the global volume can be determined as shown in Figure 6.5. Intersections between fractures are called internal intersects. Intersections between fractures and boundary surfaces are called external intersects. The middle points on intersection lengths (either internal or external intersects) are called nodes. The node of an external intersect is called an external node, while that of an internal intersect is called an internal node. The nodal representation is used to condense the fracture network to a set of nodes with one-dimensional line elements connecting these nodes.

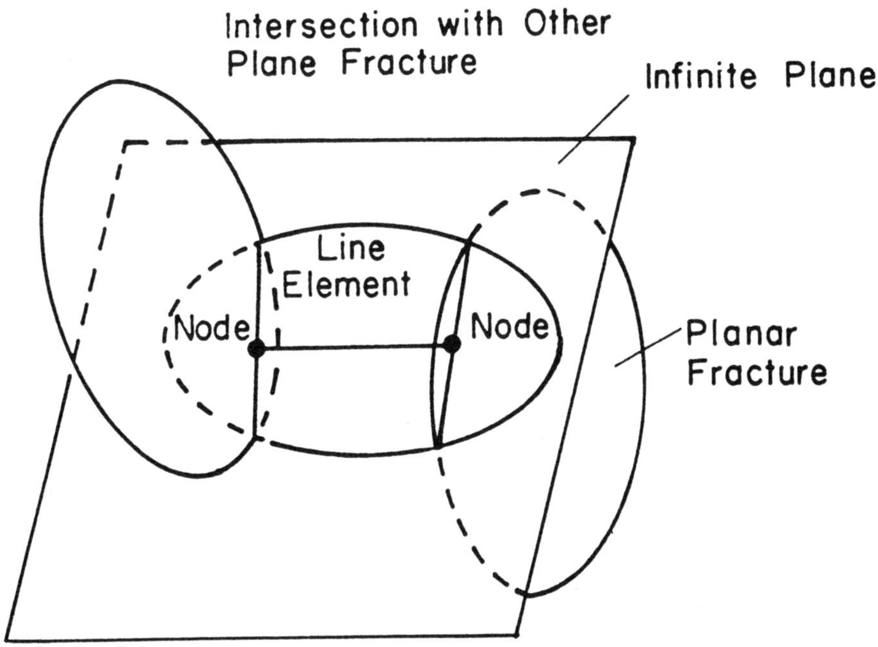

Figure 6.5 Fracture network relating model components and fractures.

Once the fracture network is assembled, the isolated fractures which cannot intersect any boundary or other fractures, and dead-end fractures which only intersect boundary surfaces and do not intersect other fractures, are removed to eliminate non-connecting flow routes. When the flow routes are found, the number of connected fractures in the flow routes can be calculated. The percolation frequency, υ, defined as the ratio of the percolation networks to the total number of fracture networks, can also be calculated. The area and perimeter of fractures inside the sample volume are then calculated. When fractures intersect the boundary surface, the area or perimeter of the fractures is calculated from the parts inside the sample volume. In this case the perimeter must be increased by adding the intersected finite length (i.e. perimeter inside sample volume plus intersected finite length). Then the average area, A_{av}, and the average perimeter, S_{av}, defined as the total area and total perimeter inside the sample volume divided by the total fractures, are also calculated. Using kinematic analysis the percolation factor is equal to (Charloix *et al.*, 1984):

$$\xi = \frac{N(A_{av})(S_{av})}{2} \tag{6.1}$$

where N is the number of fractures within the sample volume. The percolation threshold and critical percolation frequency are defined respectively as the minimal value of percolation frequency at which the fracture paths connect to all six faces of the sample volume.

To produce a well-connected medium, the fracture flow routes must be well connected to all faces in the sample volume. In other words, fractures need to connect between face 1 and face 2, between face 3 and face 4, and between face 5 and face 6 (Figure 6.3). The fracture flow route is represented as a cluster in the percolation literature.

To summarize, the discrete fracture flow model is designed to study the effect of fracture geometry on connectivity in a fracture system using statistically conditioned input. Thus, the distribution of fracture geometry may be selected from field data using a stochastic approach. Relations selected for this model are based on discontinuity analyses in Chapter 2 and include:

1. circular fracture shape
2. exponential fracture radius
3. spherical-normal fracture orientation
4. Poisson location

The computer code also allows for any number of fracture sets, each of whose parameters may belong to a unique statistic distribution.

6.3 SENSITIVITY ANALYSIS – CONNECTIVITY

Model simulation of a network of planar fractures or discontinuities in a discontinuous medium is a powerful, analytical and illustrative tool which can be used to demonstrate several important relations. Examples have been chosen specifically to illustrate:

1. the relation between the percolation factor and percolation frequency (the degree of interconnectivity);
2. the effects of fracture density and fracture radius on fracture interconnectivity; and
3. the effects of scale on the percolation threshold and the critical percolation frequency.

In the examples, the global volume is selected equal to sample volume. The mean fracture radius (L_i) and the numbers of fractures (N_i) are assumed to be the same for all sets of input parameters. Thus, three orthogonal sets will have $L_1 = L_2 = L_3$ and $N_1 = N_2 = N_3$. Orthogonal sets with or without orientation deviations are represented, so that the effects of fracture radius and fracture density on interconnectivity can be demonstrated.

Two cases are considered. Case 1 observes the effects of fracture radius and fracture density on percolation frequency and percolation factor. In this case, three scenarios (case 1.1, case 1.2 and case 1.3) are considered with known fracture sets and fracture orientation deviations. Features of each case are:

Case 1.1: Two orthogonal discontinuity sets without orientation deviations and sample volume (equal to global volume) given by 50 m × 50m × 50m.

Table 6.1 Input data for orientation in case 1 and case 2 – connectivity example

Case	Set	Orientation			
		Vertical angle A_p	Standard deviation	Horizontal angle B_p	Standard deviation
Case 1.1	Set 1	0.0	0.0	0.0	0.0
	Set 2	90.0	0.0	0.0	0.0
Case 1.2	Set 1	0.0	30.0	0.0	30.0
	Set 2	90.0	30.0	0.0	30.0
Case 1.3	Set 1	0.0	30.0	0.0	30.0
	Set 2	90.0	30.0	0.0	30.0
	set 3	0.0	30.0	90.0	30.0
Case 2.1	Set 1	0.0	30.0	0.0	30.0
	Set 2	90.0	30.0	0.0	30.0
	Set 3	0.0	30.0	90.0	30.0
Case 2.2	Set 1	0.0	30.0	0.0	30.0
	Set 2	90.0	30.0	0.0	30.0
	Set 3	0.0	30.0	90.0	30.0

Table 6.2 Input data for case 1.1, case 1.2 and case 1.3 – connectivity example

Case	Number of fractures			Fracture density (m^{-3})	Fracture radius (m)		
	Set 1	Set 2	Set 3		Set 1	Set 2	Set 3
Case 1.1	15	15	–	0.00024	10	10	–
	45	45	–	0.00072	20	20	–
	60	60	–	0.00096	40	40	–
	80	80	–	0.00128			
Case 1.2	20	20	–	0.00032	7	7	–
	60	60	–	0.00096	10	10	–
	100	100	–	0.00160	20	20	–
Case 1.3	25	25	25	0.00060	10	10	10
	40	40	40	0.00128	15	15	15
	60	60	60	0.00144	20	20	20

Table 6.3 Input data for case 2.1 and case 2.2 – connectivity example

	Number of fractures			Sample volume	Fracture density
	Set 1	Set 2	Set 3	(Global volume)	(m^{-3})
Case 2.1	16	16	16	40 m × 40 m × 40 m	0.00075
	30	30	30	50 m × 50 m × 50 m	0.00072
	52	52	52	60 m × 60 m × 60 m	0.00072
	123	123	123	80 m × 80 m × 80 m	0.00072
Case 2.2	15	15	15	25 m × 25 m × 25 m	0.00288
	30	30	30	50 m × 50 m × 50 m	0.00072
	45	45	45	75 m × 75 m × 75 m	0.00032
	75	75	75	125 m × 125 m × 125 m	0.00012

Case 1.2: Two orthogonal sets with oreintation deviations of 30° and a sample volume of 50 m × 50 m × 50 m.

Case 1.3: Three orthogonal discontinuity sets with orientation deviations of 30° and a sample volume of 50 m × 50 m × 50 m.

In each case the variables are fracture radius and fracture density. Input data are listed in Tables 6.1 and 6.2.

Case 2 observes the effect of fracture density and fracture radius on volume. In this case, two scenarios (case 2.1 and case 2.2) are considered:

Case 2.1: Three orthogonal discontinuity sets with orientation deviations of 30°. The fracture density is fixed and global volume varies.

Case 2.2: Three orthogonal sets with orientation deviations of 30°. Both fracture density and global volume vary.

Input data are listed in Tables 6.1 and 6.3.

6.3.1 EFFECT OF FRACTURE RADIUS AND DENSITY ON PERCOLATION FACTOR

The effect of fracture density on percolation factor and percolation frequency is illustrated in Figure 6.6. This plots percolation frequency against percolation factor for different numbers of fractures, equivalent to fracture density (Figure 6.6(a–c)) and different fracture radius or length (Figure 6.6(d–f)).

Percolation factor, ξ, and frequency υ have been defined in equation 5.17. The former effectively defines the ability of a discontinuous rock mass to transmit fluid in terms of the number and connectivity of its fractures. At low values below a given percolation threshold it does not have this ability; at high values flow increases rapidly. Percolation frequency effectively defines the scale dependence of flow and increases from zero with a low percolation factor to unity with a high percolation factor. The relation is illustrated in Figure 6.6 for each of the three cases.

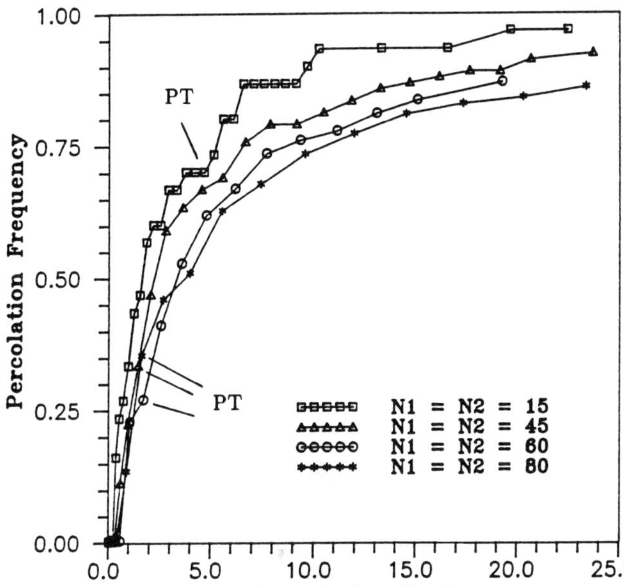

(a)

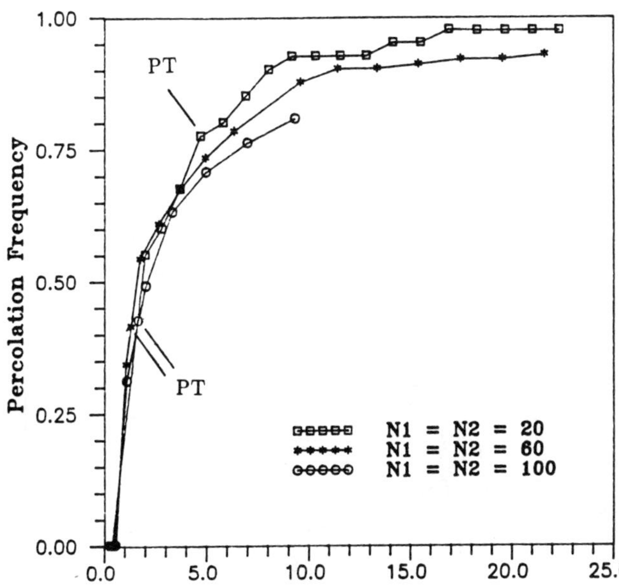

(b)

Figure 6.6 Variation of percolation frequency with percolation factor for different fracture densities (expressed as fracture numbers), for (a) case 1.1, (b) case 1.2 and (c) case 1.3; and for different fracture radii, (d) case 1.1, (e) case 1.2 and (f) case 1.3. PT denotes the percolation threshold for each curve.

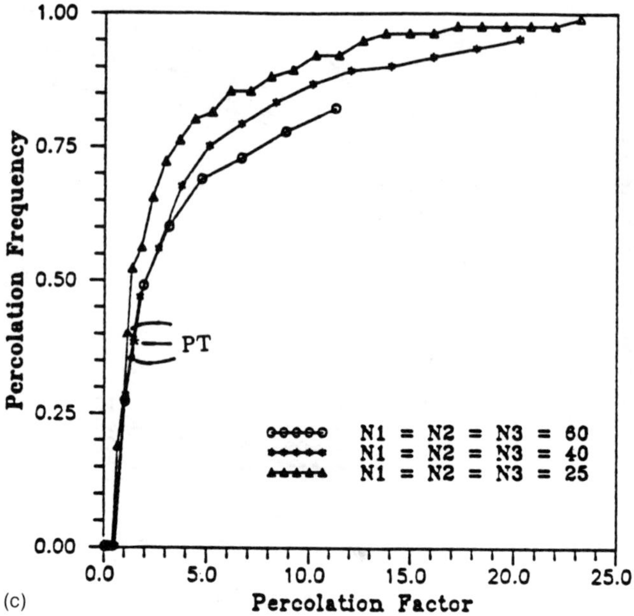

(c)

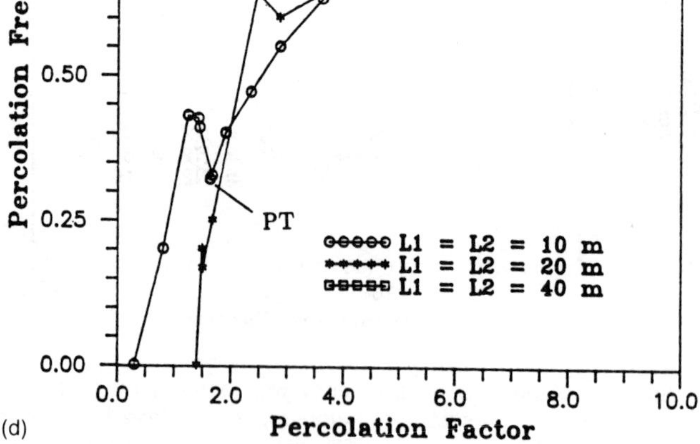

(d)

Figure 6.6 *contd.*

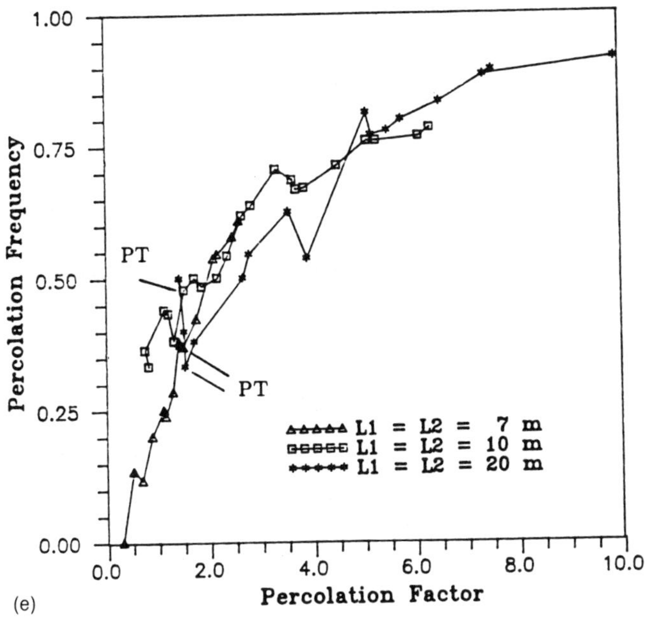

(e)

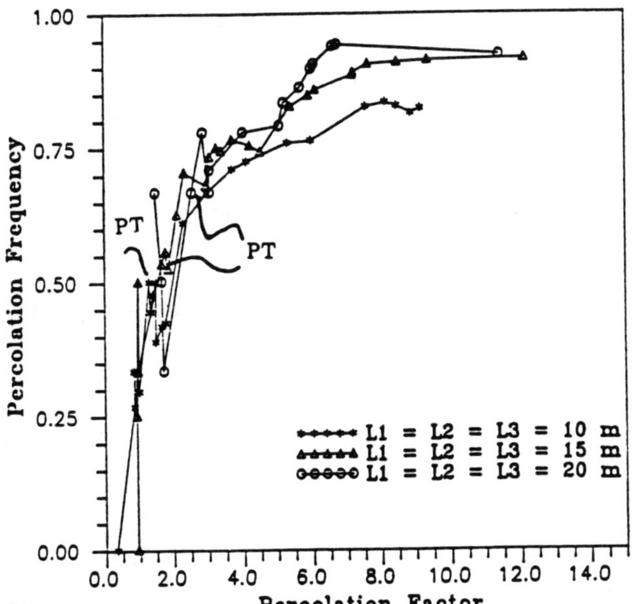

(f)

Figure 6.6 *contd.*

When a fracture system has a constant percolation factor, percolation frequencies are lower (Figure 6.6(a–c)) with a larger number of fractures (high density) than with fewer fractures (low density). This indicates that if a fracture system has a constant percolation factor, the percolation frequency of a system with low density is higher than that with a high density. The percolation frequency can be predicted to reach unity if the percolation factor becomes large or infinite. When a fracture system has a constant percolation frequency, an increase in fracture density will increase the percolation factor. The fracture system with a high fracture density will tend to be more clustered than the one with a low density. Thus, a fracture system with a high fracture density will have a higher fracture percolation factor and lower percolation frequency than one with a low fracture density. Similarly, a fracture system with low fracture density will have a lower percolation factor and higher percolation frequency than one with a high fracture density. However, this behaviour is insignificant at low percolation factors, especially below the percolation threshold. This may be due to an insufficient number of fractures or truncation of the sample boundary.

The effect of various fracture radii (or lengths) on percolation frequency is shown in Figure 6.6(d–f). At large percolation factors, fracture systems with short radius fractures generally have a lower percolation frequency than those with long radius fractures. When fracture networks have a constant percolation frequency, those with short radius fractures will tend to be more clustered than those with long radius fractures since they will have larger percolation factors. Thus, a network with large mean radius fractures will have a lower percolation factor and a higher percolation frequency than one with a small mean fracture radius, and vice versa. However, at low percolation factors, especially below the percolation threshold, the relations between percolation factor and percolation frequency appear somewhat erratic. This implies that the percolation frequency is very sensitive to fracture radius. The reason may also be insufficient fracture numbers, or possibly truncation effects at the boundary of the sampling volume.

Figure 6.7(a–c) plots percolation frequency against the number of fractures for each of the three cases and Figure 6.7(d–f) plots percolation frequency against the length or radius of fractures. The former shows that an increase in the number of fractures (i.e. increasing fracture density) increases the percolation frequency when the fracture radius is fixed. An increase in fracture density results in a decrease in fracture radius when the percolation frequency is constant.

Similarly, an increase in the fracture radius (Figure 6.7(d–f)) will increase the percolation frequency when the number of fractures is constant, and an increase in fracture radius will decrease the fracture density when the value of the percolation frequency is constant.

This suggests that if a fracture system has a higher percolation factor, especially above the percolation threshold, a network with a long radius and a low density of fractures will have a higher percolation frequency than one with a short radius and a high density of fractures. The reason for this is that a constant percolation factor in a fracture system can be maintained either by increasing the fracture density and decreasing the fracture radius, or by decreasing the fracture density and increasing the

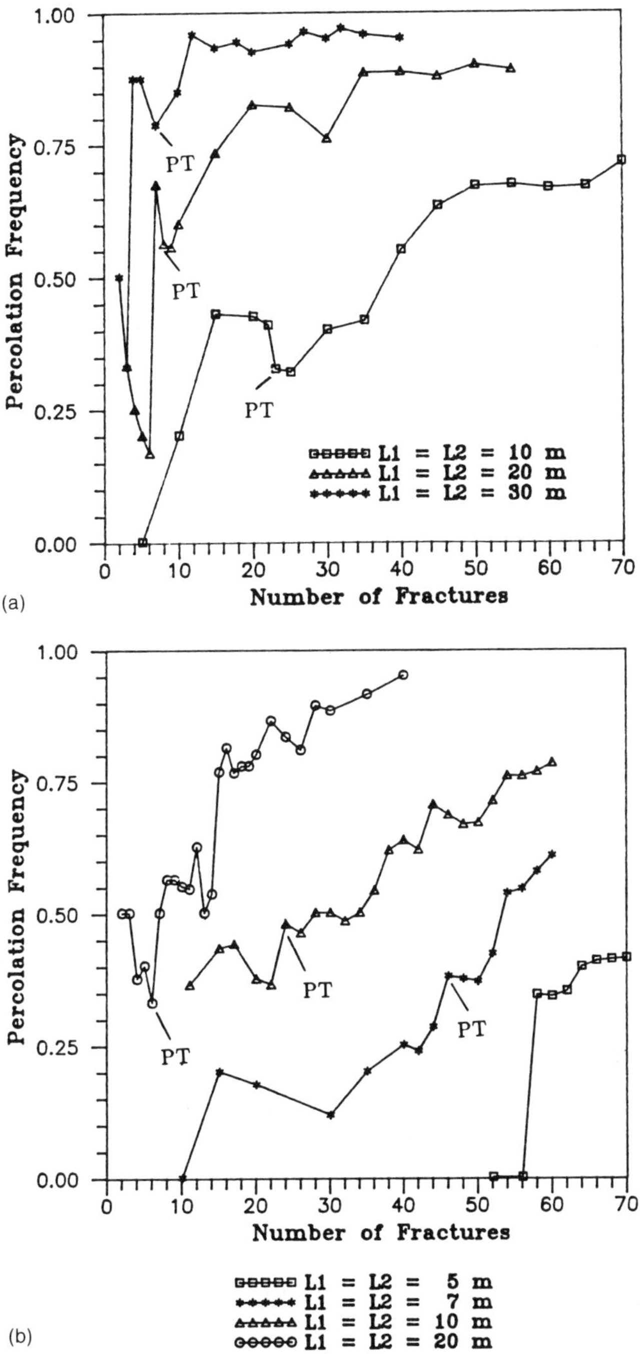
1.00
0.75
0.50
0.25
0.00
Percolation Frequency
PT
PT
PT
L1 = L2 = 10 m
L1 = L2 = 20 m
L1 = L2 = 30 m
0 10 20 30 40 50 60 70
Number of Fractures
(a)
1.00
0.75
0.50
0.25
0.00
Percolation Frequency
PT
PT
PT
PT
0 10 20 30 40 50 60 70
Number of Fractures
L1 = L2 = 5 m
L1 = L2 = 7 m
L1 = L2 = 10 m
L1 = L2 = 20 m
(b)

Figure 6.7

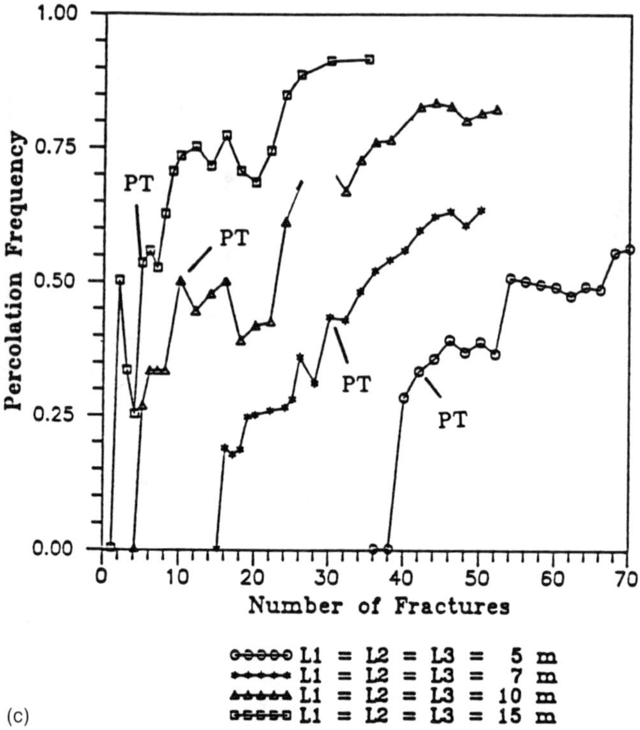

(c)

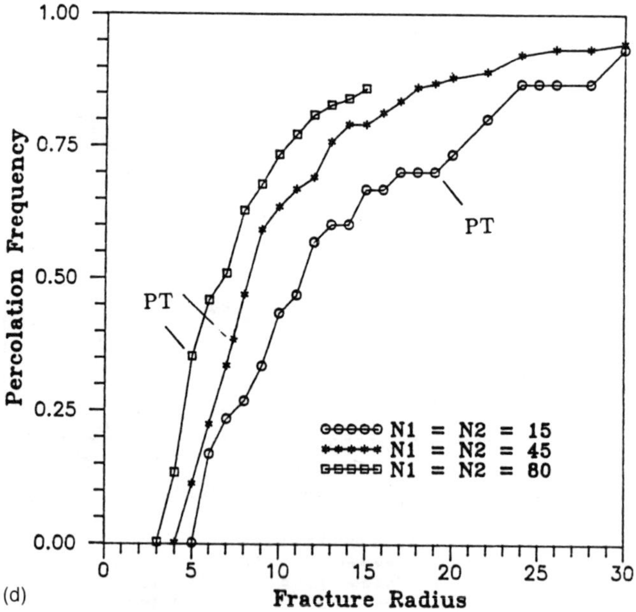

(d)

Figure 6.7 *contd.*

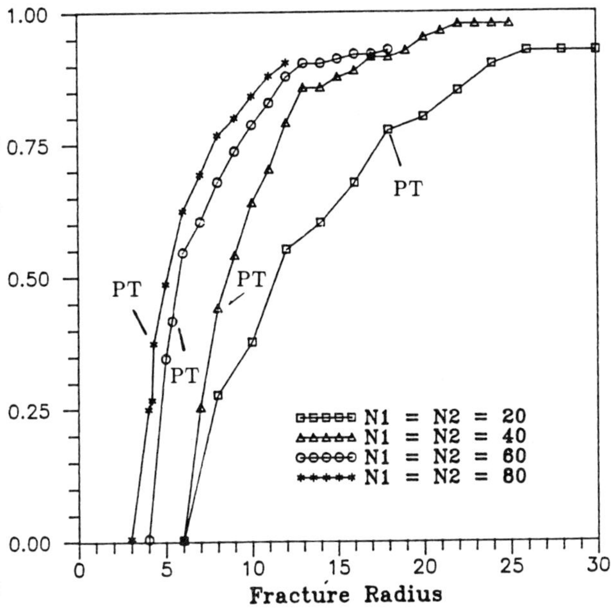

(e)

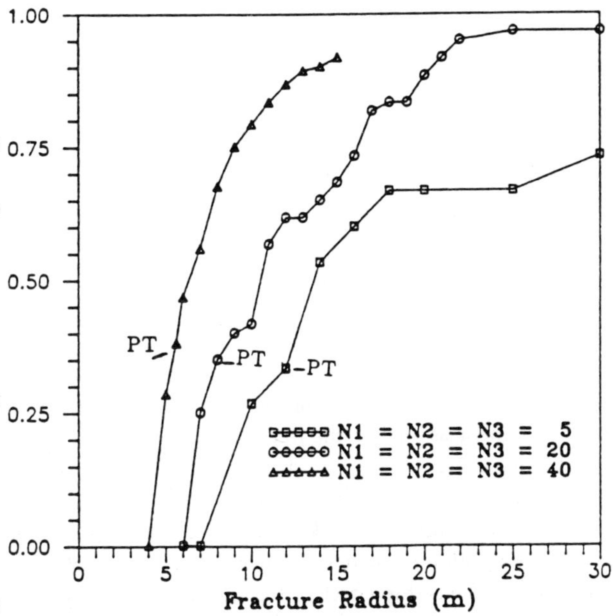

(f)

Figure 6.7 Percolation frequency as a function of fracture density for various mean fracture radii (case 1.1(a), case 1.2(b) and case 1.3(c)) and as a function of fracture radius for various frequencies (case 1.1(d), case 1.2(e) and case 1.3(f)). PT denotes the percolation threshold.

fracture radius. However, an increase in fracture density accompanied by a decrease in the fracture radius will tend to increase the connected clusters and this will lead to an increase in the percolation factor.

The simulation model can also be used to illustrate the effect of fracture radii and the number of fractures on the percolation threshold (Figure 6.8). This shows that values of percolation threshold do not vary widely or systematically with fracture density. Except for less dense fractures the computed values are similar to those of Wilke *et al.*'s (1985) percolation threshold range of 0.9–2.4. Values of percolation threshold tend to increase almost linearly with fracture radius (Figure 6.8(b)), and those for long mean radius fracture fall above Wilke *et al.*'s range of values. This implies that the percolation threshold is very sensitive to changes in fracture radius and satisfies the range of 0.9–2.4 only when the fracture radius of the system is not too long. A similar argument about fracture density is not supported by the illustration. The related critical percolation frequency falls in the range 0.22–0.56.

6.3.2 EFFECT OF SCALE ON PERCOLATION FACTOR

The relation between percolation frequency and percolation factor at different scales is illustrated in Figure 6.9 for both case 2 examples. In Figure 6.9(a) the global volume is increased with a fixed fracture density. The increase in scale at a given percolation factor results in a decrease in percolation frequency. This appears to differ from Englman *et al.*'s (1983) observation that increasing scale will lead to increases in interconnectivity. However, as has been demonstrated above, in a constant volume fracture system with a given percolation factor, an increase in mean fracture radius will lead to a decrease in fracture density. In Figure 6.9(a) fracture density is constant for each volume (Table 6.3, case 2.1). Thus the value of percolation factor is only proportional to the mean fracture radius and is independent of density. When Figure 6.9(a) is replotted in Figure 6.9(c), as percolation frequency against fracture radius, an increase in mean factors radius leads to an increase in percolation frequency. For large percolation factors, the large scale induces a large percolation frequency with constant mean fracture radius, and thus constant percolation factor. However, at low percolation factors, the percolation frequency of the larger scale is lower than that of the small scale. This result is different from Englman *et al.*'s (1983) observations and it appears that percolation factor and percolation frequency are dependent on the scale of measurement as well as the fracture radius.

In Figure 6.9(b) (case 2.2) the fracture density varies with scale. Here an increase in the percolation factor will increase the percolation frequency at a constant volume. When this is replotted as Figure 6.9(d), keeping a constant volume, a less dense rock mass has a lower percolation frequency and hence lower percolation factor although it has a larger volume. Thus, the effect of fracture density and scale on percolation frequency is significant, and the percolation factor and percolation frequency are both dependent on fracture density and scale.

The implication of this is that sample volume affects the percolation frequency if the mean fracture radius and fracture density in the field are assumed to be constant.

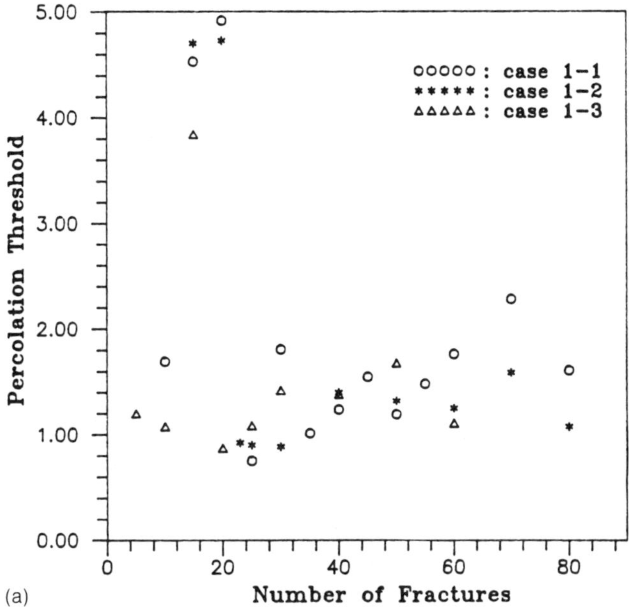

(a)

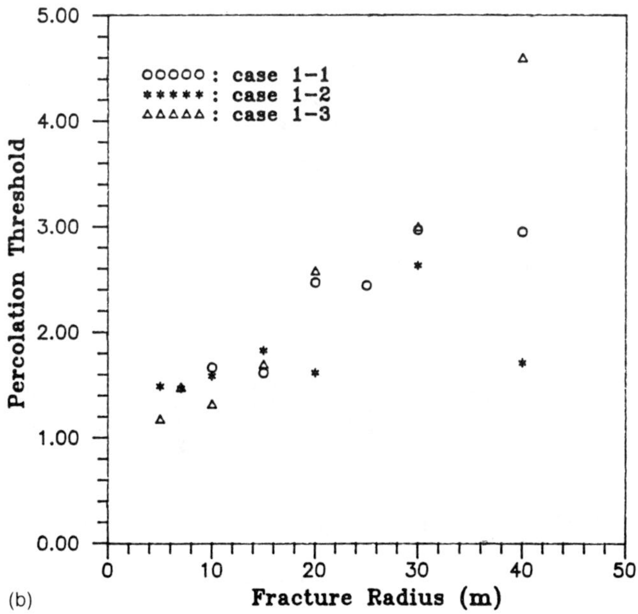

(b)

Figure 6.8 Variation in percolation threshold (a) with fracture frequency and (b) with mean fracture radius for case 1.

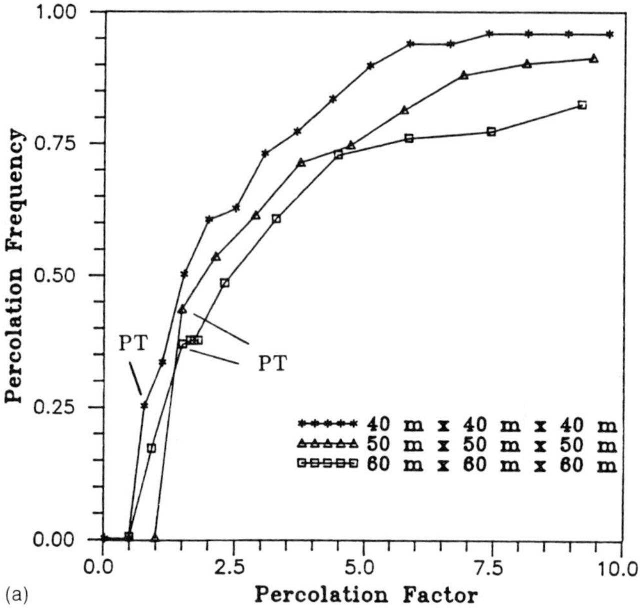

(a)

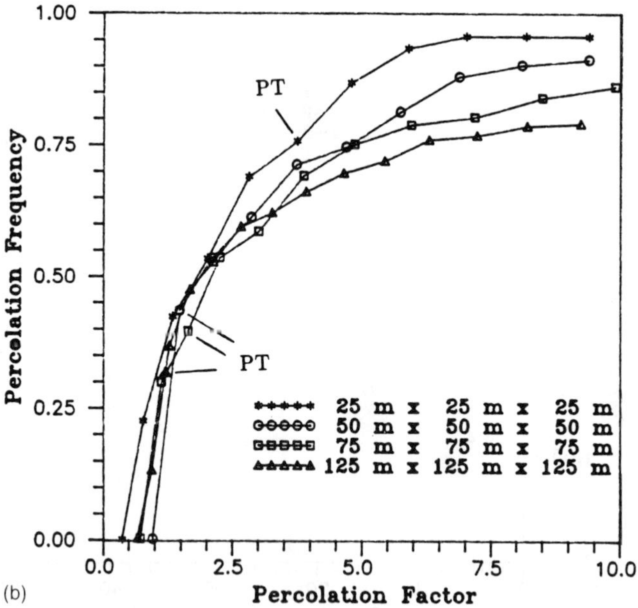

(b)

Figure 6.9 Variation in percolation frequency with percolation factor and fracture radius at different scales for constant fracture density (case 2.1, examples a and c) and variable fracture density (case 2.2, examples b and d).

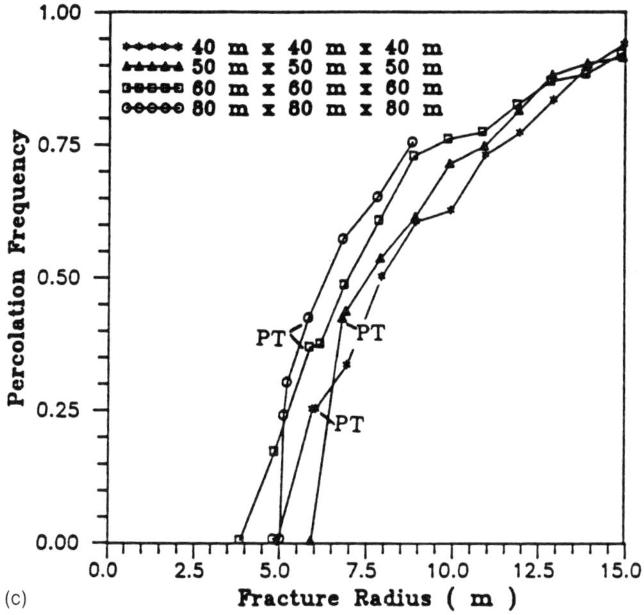

40 m x 40 m x 40 m
50 m x 50 m x 50 m
60 m x 60 m x 60 m
80 m x 80 m x 80 m
Percolation Frequency
Fracture Radius (m)
PT
PT
PT
1.00
0.75
0.50
0.25
0.00
0.0
2.5
5.0
7.5
10.0
12.5
15.0

(c)

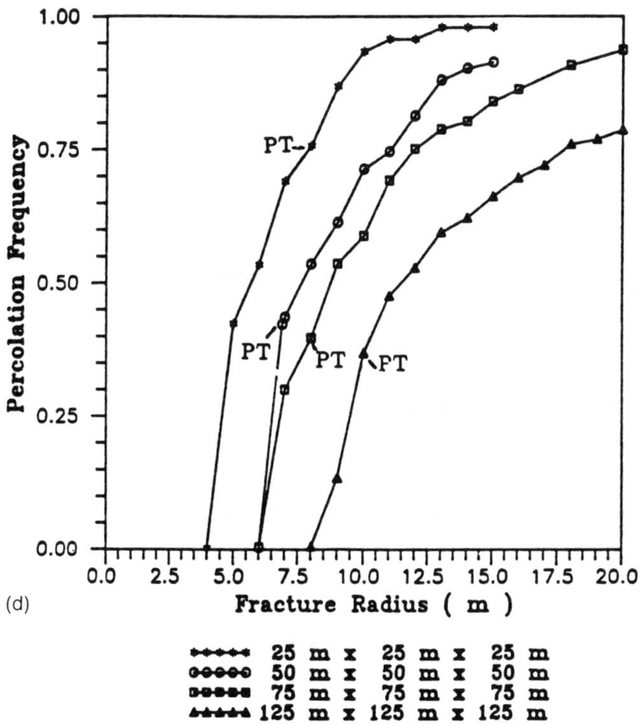

Percolation Frequency
Fracture Radius (m)
PT
PT
PT
PT
1.00
0.75
0.50
0.25
0.00
0.0
2.5
5.0
7.5
10.0
12.5
15.0
17.5
20.0
25 m x 25 m x 25 m
50 m x 50 m x 50 m
75 m x 75 m x 75 m
125 m x 125 m x 125 m

(d)

A large volume will have a higher percolation frequency than a small one when the volumes have a constant percolation factor. But a large volume will have lower percolation frequency than a small volume when the large volume has a lower percolation factor. Thus, the percolation frequency and percolation factor are dependent on fracture radius, fracture density **and** the scale of the measurement.

In order to evaluate the relation between the measurement scale or volume and the percolation threshold these are plotted in Figure 6.10. This shows that most values of percolation threshold are in the range of 1.0–1.5 except at small volumes. The related critical percolation frequency range is 0.25–0.5 except for small volumes.

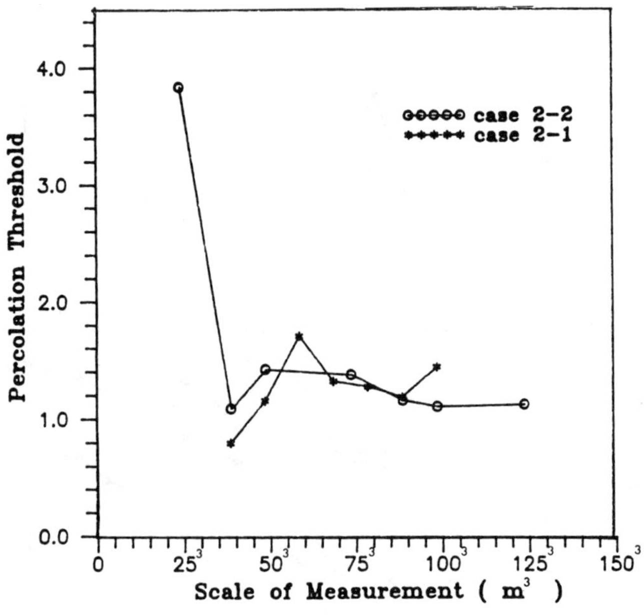

Figure 6.10 Variation in percolation threshold with scale – case 2.

6.4 CONNECTIVITY AND SCALE EFFECT

Percolation theory can provide a semi-analytical method for determining whether a given sample is large enough to be statistically representative. It can also be used to analyse the problem of interconnectivity on a finite scale, where it still has scale dependence. Using percolation theory with a percolation threshold can demonstrate that the REV concept is not always satisfied in continuum approaches to analysis of fractured rock masses. If there is a large percolation factor, especially above the percolation threshold, then the REV concept may be acceptable in the finite area, but for lower percolation factors, especially below the percolation threshold, it is not acceptable.

The simulation examples demonstrate that:

1. Above the percolation threshold, a network with a high fracture density will have a higher percolation factor and a lower percolation frequency than one with a low fracture density. A network with a large mean fracture radius will have a lower percolation factor and a higher percolation frequency than one with a short mean fracture radius. When the fracture radius is long in comparison with the sample volume, the effect of the fracture radius on the percolation frequency will be erratic. In this case, the percolation frequency will be very sensitive to fracture radius, and if the tested volume has the same fracture density and fracture radius and hence a constant percolation factor, a large volume will have a higher percolation frequency that a small one. However, if the tested volumes have the same fracture radius, a large volume with a low fracture density and hence a low percolation factor will have a lower percolation frequency than a small one with a high fracture density.
2. The percolation frequency or the degree of interconnectivity of a fracture system and the percolation factor will increase when the fracture radii and fracture density increase. However, an increase in the mean fracture radius or the fracture density within a fracture system will decrease the rate of change of percolation frequency. The rate of change will approach zero with infinite or large fracture radii or large increases in fracture density.
3. In a fracture medium with $\xi > \xi^*$, the degree of interconnectivity of short radius high density fractures will be less than that of long radius and low density fractures since the latter will have higher percolation frequency. Thus a fracture system with a lower fracture density and larger fracture radii in comparison with its volume would be expected to behave more like a porous medium.
4. If the fracture system is such that, in comparison with its volume, fracture radii are not 'too long' and fracture density is not 'too low', most values of the percolation threshold and critical percolation frequency will fall in the range of 0.9–2.4 and 0.22–0.56, respectively. In fact it appears that most values of percolation threshold and critical percolation frequency are in the range of 1.0–1.5 and 0.25–0.5, respectively.
5. The percolation threshold and critical percolation frequency are not constant values, but are dependent on scale. Thus, when the percolation threshold of a rock mass with orthogonal sets is bigger than 2.4, the rock mass can be expected to reach a connective condition. When the value is less than 0.9, the medium is non-connective. However, values between 0.9 and 2.4 must be checked repeatedly from small and large samples.

It is interesting to compare the interconnectivity index defined by Rouleau and Gale (1985) and the percolation threshold. The interconnectivity index is a way of incorporating fracture orientation, trace length and spacing into a characterization of the interconnectivity of a fracture system (section 5.2). In the fracture network simulation model used here, the shape of fractures, fracture length, Poisson orientation and the number of fractures in each set are incorporated. The advantage of this model is not only that random fractures can be generated according to the density distribution forms of fracture parameters from the field data, but also that the effect on the percolation threshold of various scales can be considered. The percolation factor can contribute more than the interconnectivity index, at least, from theoretical and statistical standpoints.

The degree of interconnectivity has been studied in two dimensions using the *LD* (length–density) concept by Long and Witherspoon (1985). They have shown that interconnectivity is dependent on fracture length and fracture density in two dimensions. In the simulation model, the percolation factor and the percolation frequency are used to describe the degree of connectivity in three dimensions and then to compare the percolation threshold to general invariants in regular or periodic lattices in order to understand the validity of the invariants. The percolation factor is defined as a function of the fracture density, fracture area and fracture perimeter, independent of fracture geometry. Basically, the results of the examples agree with those of Long and Witherspoon (1985) who state that interconnectivity is dependent on fracture length and fracture density in two dimensions.

The renormalization group approach method using Monte Carlo simulation can be used to obtain critical probability, p_c (Ma, 1973; Vicsek and Kertesz, 1981; Stauffer, 1985). Once p_c is known, the invariant can be obtained if the number of co-ordinates, Z, is known. The question is how to obtain p_c in a complex geological field and how to predict the value of Z. These are difficult to evaluate in a rock mass. It seems more practical to use an index such as the percolation factor and the percolation frequency to represent the connectivity of a three-dimensional rock mass, and to compare the percolation threshold to the invariants of regular lattices in the percolation literature.

6.5 BOUNDARY INTEGRAL ELEMENT METHOD

One numerical simulation technique which has been used to predict the flow properties of a single rigid fracture of constant aperture is the **boundary integral element method** (BIEM) (Brebbia, 1978; Patterson and Elsebai, 1982; Elsworth, 1986, 1987; Rasmussen, 1987, 1988). Since a global system may contain tens or hundreds of such fractures, the use of the boundary element formulation can converge nodes so that they are restricted to fracture edges and intersections only. The external and the interregion boundaries of the flow domain can then be discretized, and the flux along the boundary of the fracture can then be evaluated directly without further use of numerical approximations. On the other hand, the flux along the interface boundary is critical in determining the new location of the interface. Thus, for large numbers of nodes and special meshing techniques, the BIEM provides a reasonable number of nodes to converge within the fracture planes, and thus may better analyse the fracture flow problem. The basic theoretical processes are presented in the following equations.

As fluid flow percolates through a porous medium, the continuity equation for a fracture is:

$$\nabla \cdot q_f(x) = 0 \text{ in } \Omega \text{ domain} \tag{6.2}$$

where ∇ is the divergence operator, q_f is the darcian flux in a fracture plane, x is the position vector, and Ω is a two-dimensional fracture flow domain. Darcian flow and constant transmissivity can be applied thus:

$$\nabla^2 h = 0 \tag{6.3}$$

An approximation solution for fluid head and flux, which minimizes the error between the true and estimated head and flux over the flow domain, is obtained using the weighted residual formulation (Brebbia and Connor, 1984; Rasmussen, 1991) in two dimensions:

$$\int_{\Omega} [h'(x,x')\nabla^2 h(x)]\mathrm{d}\Omega = 0 \tag{6.4}$$

which can be rewritten as:

$$C(x')h(x') + \int_B h(x)q'(x,x')\mathrm{d}B = \int_B q(x)h'(x,x')\mathrm{d}B \tag{6.5}$$

where $h'(x,x')$ is the weighted residual function, which is related to the position and flow properties within the flow domain, x' is the position vector of the weighted residual function, n is the outward unit vector normal to the boundary B, $C(x')$ is a known constant resulting from the Cauchy principal value of the surface integral $q(x) = \partial h(x)/\partial n$ and $q'(x,x') = \partial h'(x,x')/\partial n$.

By discretizing the boundary into m elements, the integral term in equation 6.5 can be rewritten as:

$$C(x')h(x') + \sum_{j=1}^{m} h^j(x)q'(x,x') = \sum_{j=1}^{m} q^j(x)h'(x,x') \tag{6.6}$$

which can be reduced to:

$$[Q]\{h\} = [H]\{q\} \tag{6.7}$$

where $\{h\}$ is the head vector on the boundary, located at discrete boundary positions, $\{q\}$ is the flux normal to the boundary location and $[Q]$, $[H]$ are the boundary integral coefficient matrices between nodes i and j.

To solve the head and flux along the element, a linear interpolation of head and flux values can be used. For flow through two intersecting fractures, the internal element can be represented as the line of intersection between the fracture planes. Then the matrix combining the intersecting fracture and the rock matrix can be expressed as

$$\begin{bmatrix} Q_1 & Q_{1i} & 0 \\ & & \\ 0 & -Q_{2i} & Q_2 \end{bmatrix} \begin{Bmatrix} h_1 \\ h_i \\ h_2 \end{Bmatrix} = \begin{bmatrix} H_1 & H_{1i} & 0 \\ & & \\ 0 & H_{2i} & H_2 \end{bmatrix} \begin{Bmatrix} q_1 \\ q_i \\ q_2 \end{Bmatrix} \tag{6.8}$$

where Q_1, H_1, Q_2, H_2 are the boundary integral coefficient matrices for fractures 1 and 2, excluding the fracture interface nodes, Q_{1i}, H_{1i}, Q_{2i}, H_{2i} are the boundary integral coefficient matrices for fractures 1 and 2, including at least one fracture interface node, h_1 and h_2 are the head vectors along the external rim of fractures 1 and 2, respectively, q_1 and q_{2i} are the flux vectors along the external rim of fractures 1 and 2, respectively and q_i and h_i are the flux and head vectors at the interface between fractures 1 and 2.

Equation 6.8 can be solved by Gaussian elimination or other direct solution methods. Once the unknown head and fluxes are determined, the head at the point of intersection between fractures within the flow domain can be calculated using the following equation:

$$h_i = \sum_j (q_j Q_{ij} - h_j H_{ij}) \tag{6.9}$$

where h_i and h_j are the total head at nodes i and j, Q_{ij} and H_{ij} are boundary integral coefficients between nodes i and j and q_j is the flux normal to the boundary at node j.

6.6 DISCRETE NETWORK MODEL FOR FLOW

On the basis of the discrete fracture model in section 6.2, the fracture orientation, fracture radius, and transmissivity are generated from the fracture distribution forms of geometrical quantities. Fracture centres are also generated within the rectangular global volume which is chosen to be larger than the flow domain. A fracture centre cannot be outside this volume. The co-ordinates of the fracture centre are generated from uniform distributions in the intervals (Huang and Evans, 1985; Rasmussen, 1988). The number of fractures follows a Poisson distribution. Each fracture is checked to determine whether it intersects or is included in the flow domain. Only the intersected fractures can be kept to analyse the flow problem.

For simplification, fracture shape is chosen as either square or circular in order to investigate the effect of fracture shape on the flow rate. The fracture radius is assumed to be a negative exponential distribution. Fracture orientation is assumed to have a spherical normal orientation. The transmissivity of fractures is assumed to have a lognormal distribution with standard deviation. This algorithm makes it possible to determine whether two fractures are connected without solving for whole fractures in the actual flow domain. The interior fractures are represented with straight lines although the intersected lines are not exactly a straight line in reality.

The flow domain is contained within a rectangular volume. The boundary conditions of the volume are given by a specified head or flow as shown in Figure 6.11. For inserted nodes and elements on fractures, each fracture is first defined by eight fracture nodes inserted around the boundary of each fracture (Figure 6.12). When a fracture intersects another fracture, the line of the intersection will insert two internal nodes and one element. Such an interior intersection fracture has its own flow domain. When a fracture intersects the boundary of the volume, the inserted fracture nodes of this fracture must be projected on to the boundary of the volume (Figure 6.13). Such fracture nodes then become boundary nodes. If the internal node or boundary node is located on the boundary of the fracture, this node is separated into two

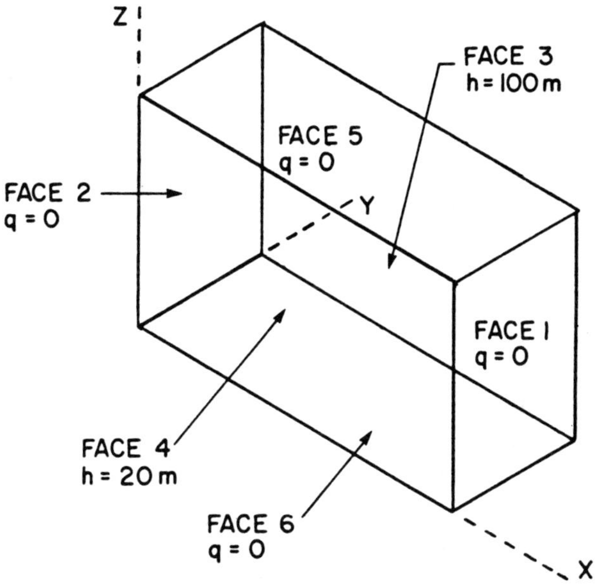

Figure 6.11 Notional flow domain for simulation. Boundary conditions are represented by a specific head of 100 m on face 3 and 20 m on face 4. All other faces are impermeable.

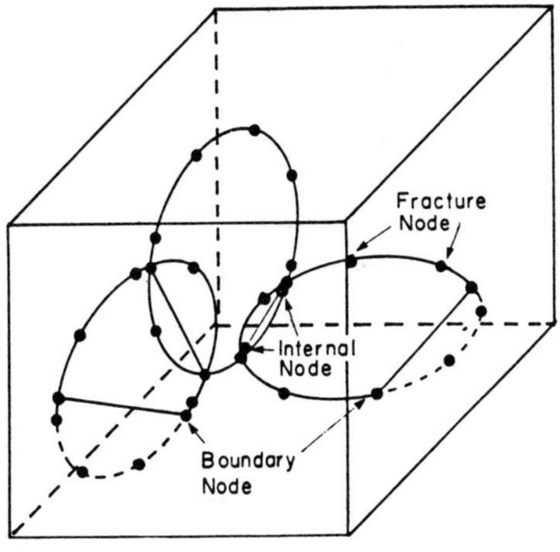

Figure 6.12 Fracture discretization scheme within a rectangular block.

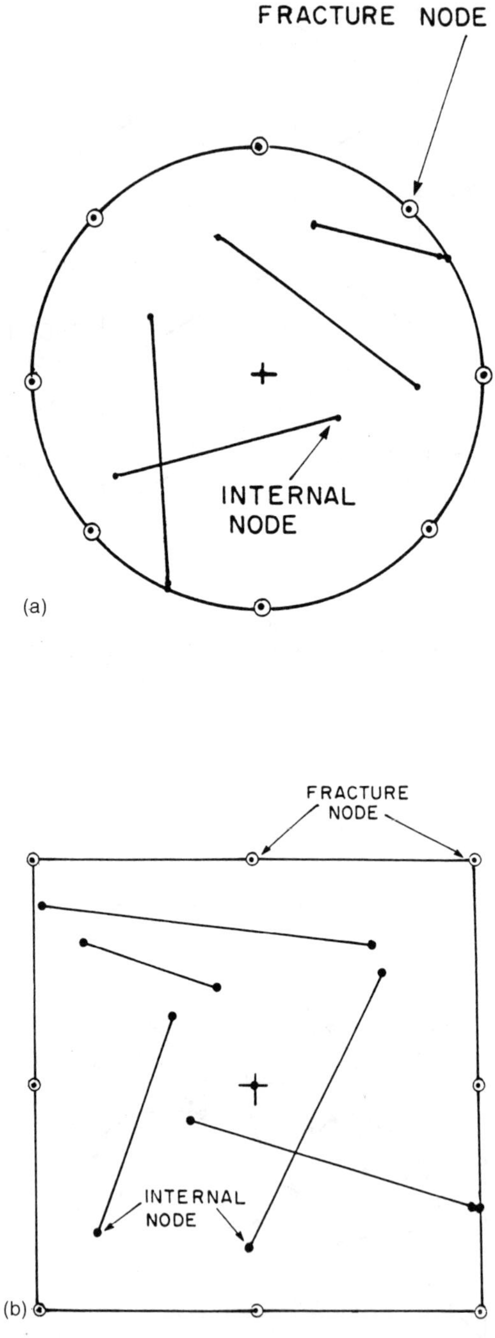

Figure 6.13 Boundary discretization scheme for (a) a circular and (b) a square fracture **not** intersecting the boundary of the block.

closely spaced nodes with one node belonging to the domain of the boundary of the fracture (i.e. fracture node) and the other node belonging to the domain of the interior fracture (either internal node or boundary node), as shown in Figure 6.14.

In Figure 6.14, the boundary of the fracture is composed of square or circular fractures which define the extent of the fracture, and straight lines, which are the intersections with the volume boundary. The straight lines inside the fracture are the intersections with other fractures. Each fracture has a uniform transmissivity. At the boundaries of the fracture, the boundary condition is given a specified head and no flow. The flow, head and flux are calculated by means of the boundary integral element method described in section 6.5. Then the hydraulic head at the interior line can be calculated. The sum of the incoming fluxes is then checked to see whether this sum approaches zero (i.e. mass balance).

6.7 SENSITIVITY ANALYSIS – FLOW

Three scenarios are presented for this case, each scenario involving a constant mean fracture radius and varying fracture density. The statistical data are specified in Table 6.4 for both circular and square shapes. The following conclusions can be made on the basis of the results presented in Table 6.4.

1. A network with long fractures and a high fracture density has a high mean flow rate.
2. The percolation frequency of conducting networks is high for large fracture radii and high densities.
3. A network with a high percolation factor has a higher mean flow rate than one with a low percolation factor.
4. There is a rather sharp threshold value above which almost all fracture networks conduct fluid and below which almost no fracture networks conduct fluid.

The values of critical percolation frequency and percolation threshold are approximately 0.35 and 1.1, respectively, for circular fractures. The percolation threshold value of 1.1 can be compared with the value of interconnectivity index reported by Rouleau and Gale (1985), who stated that, as the value of the interconnectivity index exceeds unity, the hydraulic properties become more significant.

Mean flow rate versus percolation frequency data for fractures with square and circular shapes are tabulated in Table 6.4 and plotted in Figure 6.15. Mean flow rate versus fractures per fracture set is plotted in Figure 6.16. These figures indicate that the fracture shape affects the value of the mean flow rate, the connectivity of square fractures is higher than that of circular fractures and the mean flow rate, of square fractures is higher.

In practice, estimates of the behaviour of a fracture network cannot be based solely on geometrical parameters. If it is possible to get reasonable and sufficient estimates of fracture parameters and their distributions, good quality estimates can be obtained by flow simulation. *In situ* hydraulic tests must, however, be obtained in order to check the validity of results.

The percolation factor and percolation frequency may be a good index for

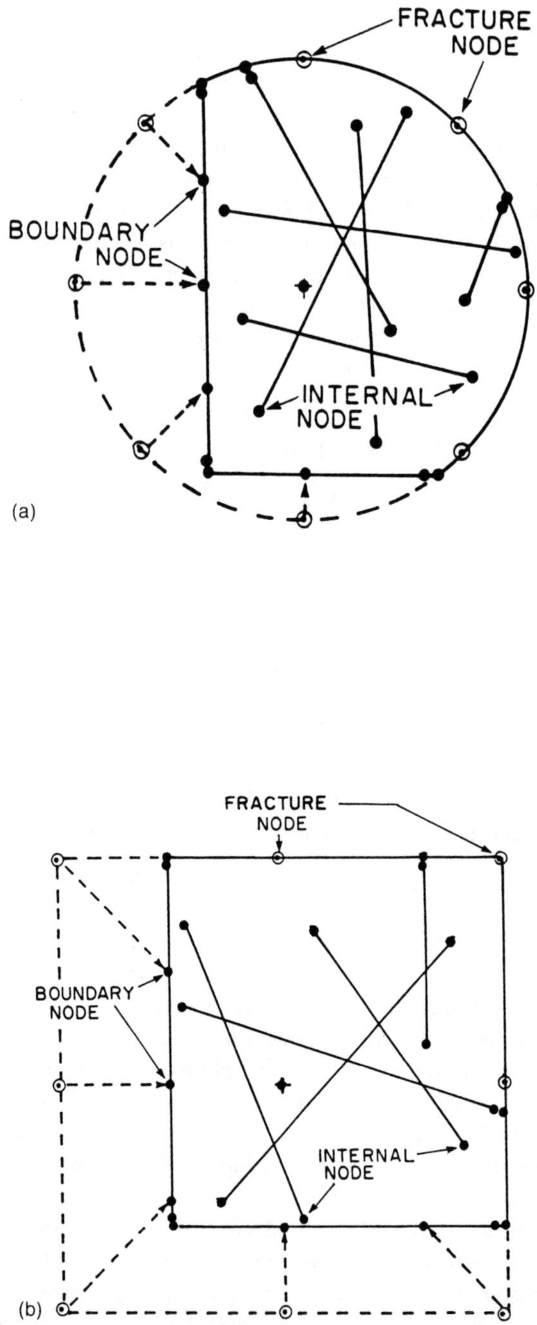

Figure 6.14 Boundary discretization scheme for (a) a circular and (b) a square fracture intersecting the boundary of the block.

Table 6.4 Input and output data of fracture parameters for flow example

Input data			Output data				
No. of fractures per set	Density per set ($10^{-3}/m^3$)	Fracture radius (m)	Circular fracture mean flow ($10^{-8}m^3/s$)	Circular fracture percolation frequency	Circular fracture percolation factor	Square fracture mean flow ($10^{-8}m^3/s$)	Square fracture percolation frequency
15	0.120	7.0	0.0*	0.178	0.555	0.0*	0.222
20	0.160	7.0	0.0*	0.250	0.698	161.8	0.333
22	0.176	7.0	0.0*	0.288	0.733	166.2	0.364
25	0.200	7.0	253.7	0.288	0.960	784.4	0.507
27	0.216	7.0	456.2	0.346	1.131	1056.0	0.519
30	0.240	7.0	774.2	0.430	1.467	–	–
10	0.080	8.5	0.0*	0.400	0.870	304.1	0.500
12	0.096	8.5	0.0*	0.333	0.870	305.8	0.417
15	0.120	8.5	0.0*	0.281	0.897	365.6	0.444
17	0.136	8.5	390.9	0.314	1.024	814.1	0.490
20	0.160	8.5	546.3	0.367	1.134	770.6	0.417
22	0.176	8.5	581.4	0.394	1.200	814.3	0.439
25	0.200	8.5	918.5	0.520	1.585	–	–
10	0.080	10.0	349.5	0.520	1.307	472.9	0.567
12	0.096	10.0	350.2	0.444	1.358	476.4	0.500
15	0.120	10.0	591.9	0.422	1.335	652.9	0.600
17	0.136	10.0	892.0	0.510	1.520	1124.4	0.608
20	0.160	10.0	814.5	0.470	1.684	1071.0	0.600
22	0.176	10.0	874.7	0.424	1.793	1158.5	0.591

*Indicates that the fracture within the sample volume is not well connected in all faces.

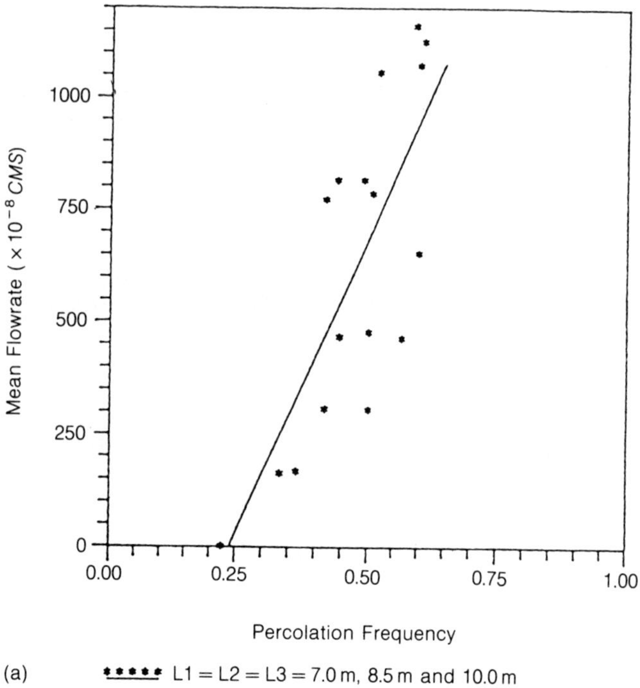

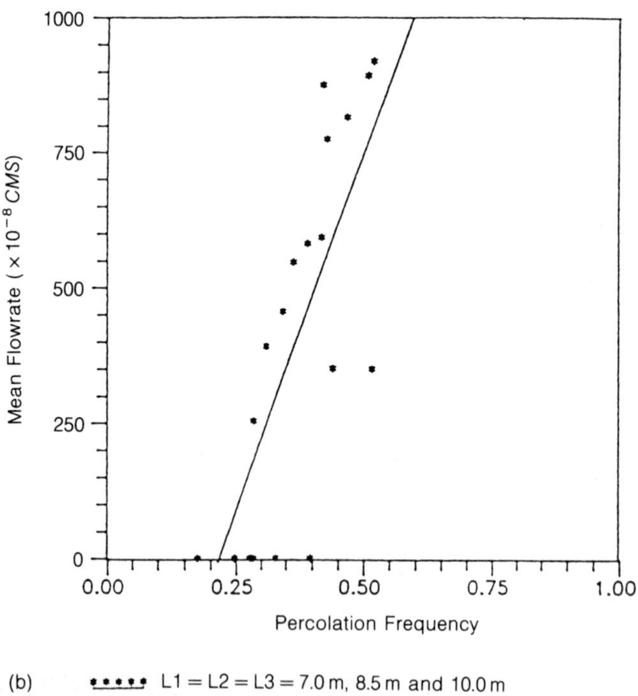

Figure 6.15 Relation between mean flow rate and percolation frequency for (a) square fractures and (b) round fractures.

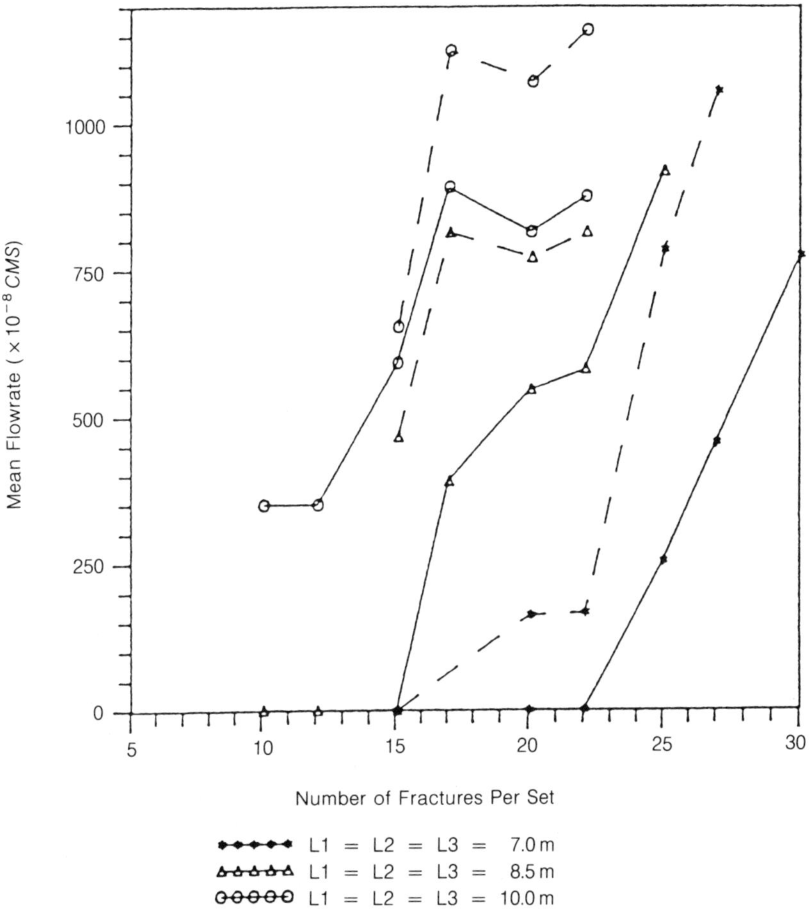

Figure 6.16 Relation between mean flow rate and fracture number for square fractures (dashed) and round fractures (solid).

determining the non-conductivity or conductivity of a network. Thus, as a first stage, fracture geometrical features should be evaluated and used to develop a percolation network. This percolation network can be used to compare with a regular lattice in percolation theory. Thus the percolation threshold (dimensional invariants) can be predicted. Using this value, it is possible to evaluate the non-conducting or conducting nature of a network in terms of scale.

Since fracture transmissivity is not generally constant, only a portion of a fracture or fractures can provide the channels which actually conduct fluid. Research into new and improved models for simulation analysis of fracture transmissivity is needed.

References

Acres Consulting Services Ltd (1978) Radioactive Waste Repository Study Part II, AECL-6188-2, Toronto, 198 pp.

Anderson, E.M. (1951) *The Dynamics of Faulting and Dyke Formation with Applications to Britain*, 2nd edn, Oliver and Boyd, Edinburgh.

Andersson, J. and Dverstorp, B. (1987) Conditional simulations of fluid flow in three-dimensional networks of discrete fractures. *Water Resources Res.*, **23**, 1876–86.

Archie, G.E. (1950) Introduction to petrophysics of reservoir rocks. *Bull. Am. Assoc. Petrol. Geologists*, **34**, 943–61.

Atchison, T.C. and Tournay, W.E. (1959) Comparative Studies of Explosives in Granite, US Bureau of Mines, RI 5509, 29 pp.

Attewell, P.B. and Farmer, I.W. (1976) *Principles of Engineering Geology*, Chapman & Hall, London.

Baecher, G.B., Lanney, N.A. and Einstein, H.H. (1977) Statistical description of rock properties and sampling. *Proc. 18th US Symp. Rock Mechanics*, pp. 501–8.

Bandis, S.C., Barton, N.R. and Christianson, M. (1985) Application of a new numerical model of joint behavior to rock mechanics problems. *Proc. Int. Symp. on Fundamentals of Rock Joints, Bjorkliden.*

Bandis, S.C., Lumsden, A.C. and Barton, N.R. (1981) Experiment studies of scale effect on the shear behavior of rock joints. *Int. J. Rock Mech. Min. Sci. & Geomech. Abstr.*, **18**, 1–20.

Bandis, S.C., Lumsden, A.C. and Barton, N.R. (1983) Fundamentals of rock joint deformation. *Int. J. Rock Mech. Min. Sci. & Geomech. Abstr.*, **20**, 249–68.

Barron, K. (1978) An air injection technique for investigating the integrity of pillars and ribs in coal mines. *Int. J. Rock Mech. Min. Sci. Geomech. Abstr.*, **15**, 69–76.

Barton, C.M. (1977) A Geotechnical Analysis of Rock Structure and Fabric in the CSA Mine, Cobor, New South Wales, CSIRO Div. Appl. Geomech. Tech. paper 24, 30–7.

Barton, N.R. (1973) Review of new shear strength criterion for rock joints. *Engineering Geol.*, **8**, 287–332.

Barton, N.R. (1976) Rock mechanics review – the shear strength of rock and rock joints. *Int. J. Rock Mech. Min. Sci. & Geomech. Abstr.*, **13**, 255–79.

Barton, N.R. and Bandis, S.C. (1980) Some effects of scale on the shear strength of joints. *Int. J. Rock Mech. Min. Sci. & Geomech. Abstr.*, **17**, 69–73.

Barton, N.R., Bandis, S.C. and Bakhtar, K. (1985) Strength, deformation and conductibility coupling of rock joints. *Int. J. Rock Mech. Min. Sci. & Geomech. Abstr.* **22**, 121–40.

Barton, N.R., Lien, R. and Lunde, J. (1974) Engineering classification of rock masses for the design of tunnel support. *Rock Mechanics*, **6**, 189–236.

Barton, N.R. and Lingle, R. (1982) Rock mass characterization methods for nuclear waste repositories in jointed rock. *Rock Mechanics: Caverns and Pressure Shafts*, ISRM Symposium, Aachen, pp. 3–18.

Bawden, W.F., Curran, J.H. and Roegiers, J.C. (1980) Influence of fracture deformation on secondary permeability – a numerical approach. *Int. J. Rock Mech. Min. Sci. & Geomech. Abstr.*, **17**, 265–79.

Bernabe, Y. (1986) The effective pressure law for permeability in Chelmsford Granite and Barre Granite. *Int. J. Rock Mech. Min. Sci. & Geomech. Abstr.*, **23**, 267–75.

Brace, W.F. (1978) A note on permeability changes in geologic materials due to stress. *Paleophysics*, **116**, 627–33.

Brace, W.F. (1980) Permeability of argillaceous and crystalline rocks. *Int. J. Rock Mech. Min. Sci. & Geomech. Abstr.*, **17**, 241–51.

Brace, W.F., Walsh, J.B. and Prangos, W.T. (1968) Permeability of granite under high pressure. *J. Geophys. Res.*, **73**, 2225–36.

Bradshaw, R.L. and McClain, W.C. (eds) (1971) Project Salt Vault – A Demonstration of the Disposal of High Activity Solidified Wastes in Underground Salt Mines, URNL-4555.

Brebbia, C.A. (1978) *The Boundary Element Method for Engineers*, Pentech Press, Plymouth.

Brebbia, C.A. and Connor, J.J. (1984) Boundary integral formulations, *Topics in Boundary Element Research*, (ed. C.A. Brebbia), **1**, pp. 1–12, Pentech Press, Plymouth.

Bridges, M.C. (1975) Presentation of fracture data for rock mechanics, *Proc. 2nd Australian–New Zealand Conf. on Geomech., Brisbane*, pp. 144–8.

Brown, E.T. (ed.) (1981) *Rock Characterization, Test and Monitoring*, Pergamon Press, Oxford.

Brown, E.T. and Hoek, E. (1978) Trends in relationships between measured in situ stresses and depth. *Int. J. Rock Mech. Min. Sci. & Geomech. Abstr.*, **15**, 211–15.

Brown, S.R. (1987) Fluid flow through rock joints: the effect of surface roughness. *J. Geophys. Res.*, **92**, 1337–47.

Brown, S.R. and Scholz, C.H. (1986) Closure of rock joints. *J. Geophys. Res.*, **91**, 4939–49.

Burgess, A. (1977) Groundwater Movements Around a Repository – Regional Groundwater Flow Analysis, KBS 54:03, 116 pp. Kaernbraenslesaekerhet, Stockholm, Sweden.

Call, R.B. (1976) Estimation of joint set characteristics from surface mapping data. *Proc. 17th US Symp. on Rock Mechanics, Salt Lake City, Utah*, **2b2**, 1–9.

Carlsson, A. and Olsson, T. (1977a) Storage in excavated rock caverns. *Proc. 1st Int. Symp. Rockstore '77, Stockholm*, **2**, 301–7.

Carlsson, A. and Olsson, T. (1977b) Water Leakage in the Foresmark Tunnel, Upland, Sweden, Sver, Geol. Unders, C734, Stockholm.

Carlsson, H. (1978) A Pilot Heater Test in the Stripa Granite, Technical Project Report No. 6, 146 pp., Swedish–American Cooperative Program on Radioactive Waste Storage in Mined Caverns in Crystalline Rock, LBL 7086 SACO6 UC-70.

Carman, P.C. (1956) *Flow of Gases Through Porous Media*, Academic Press, New York.

Cattermole, J.M. and Hanson, W.R. (1962) Geologic effects of the high explosive tests in the USGS Tunnel Area, Nevada Test Site. *US Geol. Survey Prof. Paper* 382-B, 29 pp.

Charlaix, E., Guyon, E. and Rivier, N. (1984) A criterion for percolation threshold in a random array of plates, *Solid State Communications*, **50** (11), 999–1002.

Chatterji, P.C. (1966) A study of the joint pattern and groundwater movement in Jalore Granite, Rajasthan, India. *Trans. Mineral, Geological and Metallurgical Inst. of India*, **63**, 45–53.

Cox, M.A.A. and Essam, J.W. (1976) Series expansion study of the pair connectedness in site percolation models. *J. Phys. C: Solid State Phys.*, **9**, 3985–91.

Cruden, D.M. (1977) Describing the size of discontinuities. *Int. J. Rock Mech. Min. Sci.*, **14**, 133–7.

D'Andrea, D., Fischer, R.L. and Hendrickson, A.D. (1970) Crater Scaling in Granite for Small Charges, US Bureau of Mines, Report of Investigations RI 7409.

Davis, S.N. and Turk, L.J. (1964) Optimum depth of wells in crystalling rocks. *Groundwater*, **2** (2), 6–11.

Deere, D.U. (1964) Technical description of rock cores for engineering purposes. *Rock Mechanics and Engineering Geology*, **1**, 17–32.

Deere, D.U. and Miller, R.P. (1966) Engineering Classifications and Index Properties for Intact Rock, US Air Force Weapons Laboratory, Report AFWL-TR-65-16, Kirkland, NM.

De Marsily, G. (1985) Flow and transport in fractured rock: connectivity and scale effect. *Proc. International Association of Hydrologists*, **17**, 267–78.

Den Nijs, M.P.M. (1979) A relation between the temperature exponents of the eight vertex and Q-state Potts model. *J. Phys. A: Math. Gen.*, **12**, 1857–68.

Derlich, S. (1970) Underground nuclear explosion effects in granite rock fracturing. *Proc. Symp. Eng. with Nuclear Explosives, Las Vegas, NV*, pp. 508–18.

Dershowitz, W.S. and Einstein, H.H. (1988) Characterizing rock joint geometry with joint system models. *Rock Mechanics and Rock Engineering*, **21**, 21–51.

Dienes, J.K. (1982) Permeability, percolation and statistical crack mechanics. *Proc. 23rd US Symp. on Rock Mechanics, Berkeley, CA*, pp. 86–94.

Drukovanyi, M.F. (1976) Calculation of fracture zones created by exploding cylindrical charges in ledge rocks. *Soviet Mining Science*, **12**, 292–5.

Dverstorp, B. and Andersson, J. (1989) Application of the discrete fracture network concept with field data: possibilities of model calibration and validation. *Water Resources Res.*, **25** (3), 540–50.

Einstein, H.H. and Baecher, G.B. (1983) Probabilistic and statistical methods in engineering geology – specific methods and examples. *Rock Mechanics and Rock Engineering*, **16**, 39–72.

Einstein, H.H., Baecher, G.B. and Veneziano, D. (1979) Risk Analysis for Rock Slopes in Open Pit Mines, Part I and Final Technical Report to US Bureau of Mines.

Elliott, G. M., Brown, E.T., Boodt, P.I. and Hudson, J.A. (1985) Hydromechanical behaviour of joints in the Carnmenellis Granite, S.W. England: a keynote lecture. *Proceedings of the International Symposium on Fundamentals of Rock Joints, Bjorkliden*, pp. 249–58.

Elsworth, D. (1986) A model to evaluate the transient hydraulic response of three-dimensional sparsely fractured rock masses. *Water Resources Res.*, **22**, 1809–19.

Elsworth, D. (1987) A boundary element–finite element procedure for porous and fractured media flow. *Water Resources Res.*, **23**, 551–60.

Elsworth, D. and Doe, T.W. (1986) Application of non-linear flow laws in determining rock fissure geometry from single borehole pumping tests. *Int. J. Rock Mech. Min. Sci. & Geomech. Abstr.*, **23**, 245–54.

Elsworth, D. and Goodman, R.E. (1986) Characterization of rock fissure hydraulic conductivity using idealized wall roughness profiles. *Int. J. Rock Mech. Min. Sci. & Geomech. Abstr.*, **23**, 233–43.

Englman, R., Gur, U. and Jaeger, Z. (1983) Fluid flow through a crack network in rocks. *J. of Applied Mechanics*, **50**, 707–11.

Eschbach, P.D., Stauffer, D. and Herrmann, H.J. (1980) Quoted in Vicsek, T. and Kertesz, J., 1981.

Farmer, I.W. (1983) *Engineering Behaviour of Rocks*, 2nd edn, Chapman & Hall, London.

Farmer, I.W. (1985) *Coal Mine Structures*, Chapman & Hall, London.

Fisekci, M.Y. and Barron, K. (1975) Methane pressure and flow measurements in coals and surrounding strata. *CIM Bull.*, **68**, 91–8.

Fisher, H.N. (1977) An interpretation of the pressure and flow data for the two fractures of the Los Alamos hot dry rock (HDR) geothermal system. *Proc. 18th US Symp. on Rock Mech., Golden, Colorado*, pp. 1B41–1B48.

Fisher, R.A. (1953) Dispersion on a sphere. *Proc. R. Soc. London*, Series A, **217**, 295–306.

Freeze, R.A. and Cherry, J.A. (1979) *Groundwater,* Prentice Hall, Englewood Cliffs, New Jersey.

Gale, J.E. (1982) The effects of fracture type (induced versus natural) on the stress–fracture closure permeability relationships. *Proc. 23rd US Symp. on Rock Mechanics, Berkeley, CA,* pp. 290–8.

Gale, J.E. (1985) A Numerical Field and Laboratory Study of Flow in Rocks with Deformable Fractures, PhD thesis, University of California, Berkeley.

Gale, J.E. (1987) Comparison of coupled fracture deformation and fluid flow models with direct measurements of fracture pore structure and stress-flow properties. *Proc. 28th US Symp. on Rock Mech., Tucson, Arizona,* pp. 1213–22.

Gale, J.E. and Raven, K.G. (1980) Effect of Sample Size on Stress-Permeability Relationship for Natural Fractures, Technical Information Report No. 48, LBL-11865, SAC-48, UC-70.

Gale, J.E., Rouleau, A., Witherspoon, P.A. and Wilson, C.R. (1982) Progress in the Hydrogeological Characterization of the Stripa Site, Technical Information Report No. 49, LBL-14878, SAC-49, UC-70.

Gale, J.E., Rouleau, A. and Atkison, L.C. (1985) Hydraulic properties of fractures. *Proc. 17th Int. Congress of Int. Assoc. of Hydrogeologists, Tucson, Arizona,* **17**, 1–16.

Gangi, A.F. (1978) Variation of whole and fractured porous rock permeability with confining pressure. *Int. J. Rock Mech. Min. Sci. & Geomech. Abstr.,* **15**, 249–57.

Garritty, P. (1983) Effect of Underground Construction on Geohydrology of Proximate Rocks, University of Newcastle upon Tyne, Report to BRE, contract No. PECD 7/9/075.

Ghosh, A. (1990) Role of Radial Cracks, Free Face and Natural Discontinuities of Fragmentation from Bench Blasting, PhD Thesis, University of Arizona.

Goodman, R.E. (1976) *Methods of Geological Engineering in Discontinuous Rocks,* West Publishing, New York.

Gramberg, J. and Roest, J.P.A. (1984) Cataclastic Effects in Rocksalt Laboratory and *In situ* Measurements, EUR 9258, CEC, Brussels.

Greenkorn, R.A. and Johnson, C.R. (1964) Directional permeability of heterogeneous anisotropic porous media. *J. Soc. Petrol. Engrs,* June, pp. 124–32.

Griggs, D.T. and Handin, J. (1960) Observations on fracture and a hypothesis of earthquakes, in *Rock Deformation* (eds D.R. Griggs and J. Handin), *Geol. Soc. Am. Memoir* 79, pp. 347–73.

Griswold, G. (1977) Site Selection and Evaluation Studies of the Waste Isolation Pilot Plant (WIPP), Los Medanos, N.M. SAND 77-0946, 48 pp.

Gueguen, Y. and Dienes, J. (1989) Transport properties of rocks from statistics and percolation. *J. Int. Assoc. for Math. Geology,* **21**, 1–13.

Hazen, A. (1983) Some Physical Properties of Sands and Gravels with Special Reference to Their Use in Filtration, 24th An. Rep. Mass, State Board of Health, Boston, pp. 541–56.

Heckman, R.A., Towse, D.F., Isherwood, D. *et al.* (1979) High Level Waste Repository Site Suitability Study, Status Report, NUREG/CR-0578.

Heuze, F.E. (1980) Scale effects in the determination of rock mass strength and deformability. *Rock Mechanics,* **12**, 167–92.

Hoek, E. and Brown, E.T. (1980) *Underground Excavations in Rock,* Institution of Mining and Metallurgy, London.

Hsieh, P.A. and Neuman, S.P. (1985) Field determination of the three-dimensional hydraulic conductivity tensor of anisotropic media: 1, Theory. *Water Resources Res.,* **21**, 1655–65.

Hsieh, P.A., Neuman, S.P., Stiles, G.K. and Simpson, E.S. (1985) Field determination of the three-dimensional hydraulic conductivity tensor of anisotropic media: 2, Methodology and application to fractured rocks. *Water Resources Res.,* **21**, 1667–76.

Huang, C. and Evans, D.D. (1985) A 3-dimensional Computer Model to Simulate Fluid Flow and Contaminant Transport Through a Rock Fracture System. NUREG/CR-4042, US Nuclear Regulatory Commission.

Hudson, J.A. and Priest, S.D. (1979) Discontinuities and rock mass geometry. *Int. J. Rock Mech. Min. Sci. & Geomech. Abstr.*, **15**, 133–7.

Hudson, J.A. and Priest, S.D. (1983) Discontinuity frequency in rock masses. *Int. J. Rock Mech. Min. Sci. & Geomech. Abstr.*, **20**, 73–89.

Hungr, O. and Coates, D.F. (1978) Deformability of rock joints and its relations to rock foundation settlements. *Can. Geotech. J.*, **15**, 239–49.

Hurr, R.T. and Richards, D.B. (1974) Engineering geologic, geophysical, hydrologic and rock-mechanics investigations of the Straight Creek Tunnel Site and Pilot Bore, Colorado: hydrologic investigations. *US Geol. Survey Prof. Paper* 815, pp. 78–92.

Isherwood, D. (1979) Geoscience Data Base Handbook for Modelling a Nuclear Waste Repository, Vol. 1, NUREG/CR-0912 VL, UCRL-52719, V1.

Iwai, K. (1976) Fundamental Studies of Fluid Flow Through a Single Fracture, PhD Thesis, University of California, Berkeley.

Jaeger, J.C. and Cook, N.G.W. (1979) *Fundamental of Rock Mechanics*, Chapman & Hall, London.

Jamison, D.B. and Cook, N.G.W. (1979) An analysis of the measured values for the state of stress in the earth's crust, in *Fundamentals of Rock Mechanics* (J.C. Jaeger and N.G.W. Cook), Chapman & Hall, London.

Johnson, B. (1983) Permeability of a simulated fracture as a function of normal stress. *Proc. 24th US Symp. on Rock Mechanics, Texas*, pp. 519–23.

Kazi, A. and Sen, Z. (1984) Discontinuity spacing and RQD estimates from finite length scanline. *Int. J. Rock Mech. Min. Sci. & Geomech. Abstr.*, **21**, 203–12.

Kazi, A. and Sen, Z. (1985) Volumetric RQD: an index of rock quality. *Proc. Int. Symp. on Fundamentals of Rock Joints, Bjorkliden*, pp. 95–102.

Kendorski, F.S. and Mahtab, M.A. (1976) Fracture patterns and anisotropy of San Manuel Quartz Monzanite. *Bull. Assoc. Engng. Geol.*, **13**, 23–31.

King, P.B. (1948) Geology of the Southern Guadalupe Mountains, Texas, *US Geol. Survey, Prof. Paper* 215.

Kirkpatrick, S. (1973) Percolation and conduction. *Rev. Mod. Phys.*, **45**, 574–88.

Kirsch, G. (1898) Die Theorie der Elastizitat und die Dedurfnisse der Festigkeitslehre. *Zeit. Ver. Deut. Ing.*, **42**, 797–807.

Knutson, C.F. and Buhor, B.F. (1963) Reservoir rock behaviour under moderate confining pressure. *Proc. 5th US Symp. on Rock Mechanics, University of Minnesota*, pp. 627–59.

Kozeny, J. (1927) Uber Kapillare Leitung des Wassers in Boden, Sitz de Wien. *Akad. der Wissenschaften*, **136.**

Kulatilake, P.H.S. (1985) Fitting Fisher distribution to discontinuity orientation data. *J. of Geological Education*, **33**, 266–9.

Kulatilake, P.H.S. (1988) State-of-the art in stochastic joint geometry modeling. *Proc. 29th US Symp. on Rock Mechanics, Minneapolis*, pp. 215–29.

Kulhawy, F.H. (1975) Stress–deformation properties of rock and rock discontinuities. *Engng Geol.*, **8**, 327–50.

Kunkel, J.R., Way, S.C. and McKee, C.R. (1988) Comparative Evaluation of Selected Continuum and Discrete-Fracture Models, US Nuclear Regulatory Commission, NUREG/CR-5240.

Lai, C.S. (1971) Fluid Flow Through Rock Salt Under Various Stress States, PhD Thesis, Michigan State University.

Laslett, G.M. (1982) Censoring and edge effects in areal and line transect sampling of rock traces. *J. Int. Assoc. Math. Geol.*, **14**, 125–40.

Lee, C.H. and Farmer, I.W. (1990a) Estimation of conducting aperture, fracture porosity and fracture permeability of rock masses. *Proc. 31st US Symp. on Rock Mech.*, pp 355–62.

Lee, C.H. and Farmer, I.W. (1990b) A simple method of estimating rock mass porosity and permeability. *Int. J. Mining and Geological Engineering*, **8**, 83–92.

Lewis, D.C. and Burgy, R.H. (1964) Hydraulic characteristics of fractured and jointed rock. *Groundwater*, **2**(3), 4–14.

Lindblom, U.E., Gnirk, P.F. and Charlwood, R.G. (1977) Groundwater Movements Around a Repository – Final Report, KBS-TR-54:06 (Sweden), 127 pp.

Londe, P. (1973) The role of rock mechanics in the reconnaissance of rock foundations. *Q. J. Engineering Geol.*, **6**, 57–64.

Long, J.C.S. (1985) Verification and Characterization of Continuum Behavior of Fractured Rock at AECL Underground Research Laboratory, Technical Report, BMI/OCRD-17. LBL-14975.

Long, J.C.S. and Witherspoon, P.A. (1985) The relationship of the degree of interconnection to permeability of fracture networks. *J. Geophys. Res.*, **90**(B4), 3087–98.

Long, J.C.S., Remer, J.S., Wilson, C.R. and Witherspoon, P.A. (1982) Porous media equivalents for networks of discontinuous fractures. *Water Resources Res.*, **18**, 645–58.

Louis, C.A. (1969) A Study of Groundwater Flow in Jointed Rock and Its Influence on the Stability of Rock Masses, Rock Mechanics Research Report No. 10, Imperial College, London.

Louis, C.A. (1974) Rock hydraulics, in *Rock Mechanics* (ed. L. Muller), Springer Verlag, Vienna, pp. 299–382.

Lundstrom, L. and Stille, H. (1978) Large Scale Permeability Test of the Granite in the Stripa Mine and Thermal Conductivity Test, Technical Project Report No. 2, 33 pp. Swedish–American Cooperative Program on Radioactive Waste Storage in Mined Caverns in Crystalling Rock, LBL 7052, SAC-02, UC-70, TID-4500-R66.

Ma, S.K. (1973) Introduction to the renormalization group. *Rev. Mod. Phys.*, **45** (4), 589–614.

Madden, T.R. (1983) Microcrack connectivity in rocks: a renormalization group approach to the critical phenomena of conduction and failure in crystalline Rocks. *J. Geophy. Res.*, **88** (B1), 585–92.

Mahtab, M.A. and Yegulalp, T.M. (1984) A similarity test for grouping orientation data in rock mechanics. *Proc. 25th US Symp. on Rock Mech., Evanston, IL*, pp. 495–502.

Mahtab, M.A., Bolstad, D.D., Alldredge, J.R. and Shanley, R.J. (1972) Analysis of Fracture Orientations for Input to Structural Models of Discontinuous rock, USB Mines RI 7669.

Maini, Y.N.T., Noorishad, J. and Sharp, J. (1972) Theoretical and field considerations on the determination of *in situ* hydraulic parameters in fractured rock. *Proc. Symp. on Percolation Through Fissured Rock*, Int. Soc. for Rock Mech. and Int. Assoc. of Eng. Geology.

Makurat, A. (1985) The effect of shear displacement on the permeability of natural rough joints. *Proc. 17th Int. Cong. of Int. Assoc. of Hydrologists*, **17**, 99–106.

Mandelbrot, B.B. (1982) *The Fractal Geometry of Nature*, Wilt Freeman, San Francisco.

Manev, G. and Avramova-Tacheva, E. (1970) On the valuation of strength and resistance condition of the rocks in natural rock massif. *Proc. 2nd Cong. Int. Soc. Rock Mech., Belgrade*, Paper 1–10, pp. 59–64.

Massachusetts Institute of Technology (1979) Risk Analysis for Rock Slopes in Open Pit Mines, Final Technical Report, Appendix UM, User's Manuals for Computer Programs, US Bureau of Mines Contract J0275015.

McCrae, R.W. (1982) Geotechnical Factors Affecting the Design of Highway Tunnels in Hard Rock, PhD Thesis, University of Newcastle upon Tyne.

McMahon, T.R. (1974) Design of rock slopes against sliding on pre-existing fractures. *Proc. 3rd Cong. Int. Soc. Rock Mechanics, Denver*, **2**, 803–8.

Miller, C.H., Cunningham, D.R. and Cunningham, M.J. (1974) An air injection technique to study the intensity of fracturing around a tunnel in volcanic rocks. *Bull. Int. Assoc. Engng Geol.*, **9**, 203–17.

Miller, S.M. (1978) The Feasibility of Applying Geostatistics to Rock Fracture Spacing. MS thesis, University of Arizona.

Mitchell, J.K. (1976) *Fundamentals of Soil Behaviour*, J. Wiley, New York.

Mojtabai, N., Cetintas, A., Farmer, I.W. and Savely, J. (1989) In-place and excavated block size distributions. *Proc. 30th US Symp. Rock Mechanics, West Virginia University*, pp. 537–44.

Montazer, P., Chitombo, G., King, R. and Ubber, W. (1982) Spatial distribution of permeability around CSM.ONWI Room, Edgar Mine, Idaho Springs, Colorado. *Proc. 23rd Symp. on Rock Mechanics, University of California, Berkeley*, pp. 47–56.

Mordecai, M. and Morris, L.H. (1970) An investigation into the changes of permeability occurring in a sandstone when failed under triaxial stress conditions. *Proc. 12th US Symp. Rock Mechanics, Rolla*, p. 221.

Nelson, R.A. and Handin, J. (1977) Experimental study of fracture permeability in porous rock. *Bull. Am. Assoc. Pet. Geol.*, **61**, 227–36.

Neuman, S.P. (1987) Stochastic continuum representation of fractured rock permeability as an alternative to REV and fracture network concepts. *Proc. 28th US Symp. on Rock Mechanics, Tucson, Arizona*, pp. 533–61.

Neuzil, C.E. and Tracy, J.V. (1981) Flow through fractures. *Water Resources Res.*, **17**, 191–9.

Nightingale, M.P. and Bolde, H.W. (1980) Quoted in Vicksek, T. and Kertesz, J. (1981).

Oda, M. (1985) Permeability tensor for discontinuous rock masses. *Géotechnique*, **35**, 483–95.

Oda, M. (1986) An equivalent continuum model for coupled stress and fluid flow analysis in jointed rock masses. *Water Resources Res.*, **22**, 1845–56.

Oda, M. and Hatsuyama, Y. (1985) Permeability tensor for jointed rock masses. *Proc. Int. Symp. on Fundamentals of Rock Joints, Bjorkliden.*

Office of Nuclear Waste Isolation (ONWI) (1978) Technical Support for CEIS: Radioactive Waste Isolation in Geologic Formations, Y/ONWI/TM-36.

Ohle, E.L. (1951) The influence of permeability on ore distribution in limestone and dolomite. *Econ. Geol.*, **46**, 667.

Olson, J.J., Willard, R.J., Fogelson, D.E. and Hjelmstad, K.E. (1973) Rock Damage from Small Charge Blasting in Granite, US. Bureau of Mines, R17751, 44 pp.

Pahl, P.J. (1981) Estimating the mean length of discontinuity trace. *Int. J. Rock Mech. Min. Sci. & Geomech. Abstr.*, **18**, 221–8.

Palmstrom, A. (1982) The volumetric joint count: a useful and simple measure of the degree of rock mass jointing. *Proc. 4th Cong. Int. Assoc. of Engng Geol.*, **5**, 221–8.

Panel Discussion (1985) In *Proc. Memoirs of the 17th Int. Cong. of Int. Assoc. of Hydrogeologists*, pp. 808–49, Int. Assoc. of Hydrogeologists, Tucson, AZ.

Patterson, C. and Elsebai, N.A.S. (1982) A regular boundary method using non-conforming elements for potential problem in three dimensions. *Boundary Element Method in Engineering*, Proc. 4th International Seminar, Southampton, England, September (ed. C.A. Brebbia), pp. 112–26.

Patton, F.D. and Deere, D.U. (1970) Significant geologic factors in rock slope stability. *Proc. Symp. Planning Open Pit Mines*, Balkema, Rotterdam, pp. 143–51.

Poole, R.W. and Farmer, I.W. (1980) Consistency and repeatability of Schmidt Hammer rebound data during field testing. *Int. J. Rock Mech. Min. Sci.*, **17**, 167–72.

Poulton, M.M., Mojtabai, N. and Farmer, I.W. (1990) Scale invariant behavior of massive and fragmented rock. *Int. J. Rock Mech. Min. Sci. & Geomech. Abstr.*, **27**, 219–21.

Pratt, H.R., Black, A.D., Brace, W.F. *et al.* (1977) Elastic and transport properties of an *in situ* jointed granite. *Int. J. Rock Mech. Min. Sci. & Geomech. Abstr.* **14**, 35–45.

Price, N. (1966) *Fault and Joint Development in Brittle and Semi-Brittle Rock*, Pergamon Press, Oxford.

Priest, S.D. (1985) *Hemispherical Projection Methods in Rock Mechanics*, George Allen Unwin, London.

Priest, S.D. and Hudson, J.A. (1976) Discontinuity spacing in rock, *Int. J. Rock Mech. & Min. Sci.*, **13**, 135–48.

Rasmussen, T.C. (1987) Computer simulation of steady fluid flow and solute transport through three-dimensional networks of variably saturated discrete fractures, in *Flow and Transport Through Unsaturated Fractured Rock* (eds D.D. Evans and T.J. Nicholson), AGU Geophysical Monograph 39, pp. 107–14.

Rasmussen, T.C. (1988) Fluid Flow and Solute Transport Through Three-Dimensional Networks of Variably Saturated Discrete Fractures, PhD Dissertation, Department of Hydrology and Water Resources, University of Arizona.

Rasmussen, T.C. (1991) Steady fluid flow and travel times in partially saturated fractures using a discrete air interface. *Water Resources Research*, **27**, 67–76.

Rasmussen, T.C., Huang, C.H. and Evans, D.D. (1985) Numerical experiments on artifically-generated, three-dimensional fracture networks: an examination of scale and aggregation effects, *Int. Assoc. of Hydrogeologists, Mem.*, **17**, Hydrogeology of Rocks of Low Permeability, pp. 676–80.

Ratigan, J.L. (1977) Groundwater Movements Around a Repository – Rock Mechanics Analysis, KBS-TR-5404 (Sweden), 50 pp.

Ratigan, J.L., Burgess, A.S., Skiba, E.L. and Charlwood, R.G. (1977) Groundwater Movements Around a Repository – Repository Domain Groundwater Flow Analysis, KBS-TR-54:05 (Sweden), 127 pp.

Raven, K.G. and Gale, J.E. (1976) Evaluation of structural and groundwater conditions in underground mines and excavations: subsurface containment of solid radioactive wastes. *Geol. Survey of Canada, Progress Report* EMR/GSC-RW.

Raven, K.G. and Gale, J.E. (1977) Subsurface Containment of Solid Radioactive Waste: A Study of the Surface and Subsurface Structural Groundwater Conditions at Selected Underground Mines and Excavations, Div. of Energy and Mines Res., EMR/GSC-RW Int. Report No. 1/77. Ottawa, 105 pp.

Raven, K.G. and Gale, J.E. (1985) Water Flow in a Natural Rock Fracture as a Function of Stress and Sample Size, *Int. J. Rock Mech. Min. Sci. & Geomech. Abstr.*, **22**, 251–61.

Reeves, M.J. (1977) Groundwater flow in early Mesozoic and Palaezoic aquifers, *Proc. Conf. on Rock Engineering, Newcastle upon Tyne*, British Geotechnical Soc., pp. 441–56.

Reynolds, T.D. (1960) Reactor Fuel Waste Disposal Project: Permeability of Rock Salt and Creep of Underground Salt Cavities, Final Report AEC Contract AT-(11-1)-490.

Robertson, A. (1970) The interpretation of geological factors for use in slope theory, in *Planning Open Pit Mines* (ed. P.W.J. van Rensburg), Balkema, Cape Town, South Africa, pp. 55–71.

Robinson, P.C. (1982) Connectivity of Fracture Systems – A Percolation Theory Approach, Theoretical Physics Division, AERE Harwell, DOE Report No. DOE/RW/81. 028, March.

Rocha, M. (1965) Some Problems on Failure of Rock Mass, Lab. Nat. Eng. Civ., Lisbon, Mem. 258.

Rocha, M. (1974) Recent possibilities of studying foundations of concrete dams, *Proc. 3rd Cong. Int. Soc. Rock Mech., Denver*, **1A**, 879–97.

Rocha, M. and Francis, F. (1977) Determination of permeability in anisotropic rock masses from integral samples. *Rock Mechanics*, **9**, 67–93.

Romm, E.S. (1966) Flow Characteristics of Fractured Rock (in Russian), Nedra, Moscow. Quoted in Ziegler, T.W., 1976.

Ross, R.E. (1945) An investigation into the laws of flow of fluids through beds of granular materials. *Proc. Instn Mech. Engrs*, **153**, 141–8.

Rouleau, A. and Gale, J.E. (1985) Statistical characterization of the fracture system in the Stripa Granite, Sweden. *Int. J. Rock Mech. Min. Sci. & Geomech. Abstr.*, **22**, 353–67.

Ryan, T.M. (1987) A Laboratory Assessment of Flow Characteristics and Permeability of Fractures in Rock, MS Thesis, University of Arizona, Tucson.

Ryan, T.M., Farmer, I.W. and Kimbrell, A.F. (1987) Laboratory determination of fracture permeability. *Proc. 28th US Symp. on Rock Mechanics, Tucson, Arizona,* pp. 595–600.

Sagar, B. and Runchal, A. (1982) Permeability of fractured rock: effect of fracture size and data uncertainties. *Water Resources Res.,* **18** (2), 266–74.

Savely, J. (1972) Orientation and Engineering Properties of Jointing in the Sierrita Pit, Arizona, MS Thesis, University of Arizona.

Scott, J.H., Lee, F.T., Carroll, R.D. and Robinson, C.S. (1968) The construction parameters in the Straight Creek Tunnel pilot bore. *Int. J. Rock Mech. Min. Sci.,* **5**, 1–30.

Sen, Z. and Kazi, A. (1984) Discontinuity spacing and RQD estimates from finite length scanlines. *Int. J. Rock. Mech. Min. Sci. & Geomech. Abstr.,* **21**, 203–12.

Shante, V.K.S. and Kirkpatrick, S. (1971) Introduction to percolation theory. *Adv. Phys.,* **20**, 325–57.

Shapiro, A. and Andersson, J. (1983) Steady state fluid response in fractured rock: a boundary element solution for a coupled, discrete fracture continuum model. *Water Resources Res.,* **19** (4), 959–69.

Sharp, J.S. and Maini, Y.N.T. (1972) Fundamental considerations on the hydraulic characteristics of joints in rock. *Proc. Symp. on Percolation Through Fissured Rock,* Int. Soc. for Rock Mech. and Int. Assoc. of Eng. Geology, T1-F.

Shehata, W.M. (1971) Geohydrology of Mount Vernon Canyon Area, Jefferson County, Colorado, PhD Thesis, Colorado School of Mines, Golden, Colorado.

Siskind, D.E. and Fumarti, R.R. (1974) Blast Produced Fractures in Lithonia Granite, US Bureau of Mines, RI 7901, 44 pp.

Siskind, D.E., Steckley, R.C. and Olson, J.J. (1973) Fracturing in the Zone Around a Blasthole, White Pine, Michigan, US Bureau of Mines, RI 7753, 20 pp.

Slichter, C.S. (1989) Theoretical investigations of the motion of groundwater. *US Geological Survey,* 19th An. Rep. Pt 2, pp. 295–384, Washington, DC.

Smith, W.O. (1933) Capillary flow through ideal uniform soil. *Physics,* **3**, 139–46.

Snow, D.T. (1965) A Parallel Plate Model of Permeable Fractured Media, PhD Thesis, University of California, Berkeley, 331 pp.

Snow D.T. (1968a) Rock fracture spacings, openings and porosities. *J. Soil. Mech. Found. Div.,* ASCE, **94** (SM1), 73–91.

Snow, D.T. (1968b) Hydraulic characteristics of fractured metamorphic rocks of the front range and implications to the Rocky Mountain Arsenal Well. *Col. Sch. Mines Quart.,* **63**, 167–99.

Snow, D.T. (1968c) Fracture deformation and changes of permeability and storage upon changes of fluid pressure. *Col. Sch. Mines Quart.,* **63**, 201–44.

Snow, D.T. (1969) Anisotropic permeability of fractured media. *Water Resources Res.,* **5**, 1273–89.

Snow, D.T. (1970) The frequency and apertures of fractures in rock. *Int. J. Rock. Mech. Min. Sci.,* **7**, 23–40.

Stauffer, D. (1985) *Introduction to Percolation Theory,* Taylor & Francis, London and Philadelphia.

Steffen, O. (1975) Recent developments in the interpretation of data from joint surveys in rock masses. *Proc. 6th Reg. Cong. for Africa on Soil Mech. and Found.,* **9**(2), 7–26.

Stewart, J.W. (1962) Water-Yielding Potential of Weathered Crystalline Rocks at the Georgia Nuclear Laboratory, US Geol. Survey Prof. Paper 450-B.

Stinchcombe, R.B. and Watson, B.P. (1976) Renormalization group approach for percolation conductivity. *J. Phys. C: Solid State Phys.,* **9**, 3221–47.

Stormont, J.C. (1990) Discontinuous behavior near excavations in a bedded salt formation. *Int. J. Mining and Geol. Eng.,* **8**, 35–56.

Sutherland, H.J. and Cave, S. (1978) Gas Permeability of SENM Rock Salt, Sandia Laboratories, Albuquerque, New Mexico, SAND-78-2287.

Swan, G. (1980) Stiffness and associated joint properties of rock. *Proc. of the Conference on the Application of Rock Mechanics to Cut and Fill Mining*, pp. 169–78.

Swan, G. (1983) Determination of stiffness and other joint properties from roughness measurements. *Rock Mechanics and Rock Engineering*, **18**, 19–38.

Terzaghi, K. (1925) Principles of soil mechanics. *Engng. News Record*, **95**, 832.

Terzaghi, R. (1965) Sources of error in joint surveys. *Géotechnique*, **15**, 287–303.

Thomas, H.E. (1974) Air injection to evaluate mine rock fractures. *Compress. Air Mag.*, **79**, 12–13.

Thorpe, R.K., Watkins, D.J. and Ralph, W.E. (1983) Strength and permeability of an extra-large specimen of granitic rock. *Proc. 24th US Symp. on Rock Mech., Texas*, pp. 511–18.

Tiren, S., Ahblom, K. and Strahle, A. (1985) Investigations of fracture zones in the Brandan Area, Northern Uppland, Sweden: geological and hydrogeological characterization. *Proc. Int. Symp. on Fundamentals of Rock Joints, Bjorkliden*, pp. 57–64.

Tsang, Y.W. (1984) The effect of tortuosity on fluid flow through a single fracture. *Water Resources Res.*, **20**(9), 1209–15.

Tsang, Y.W. and Witherspoon, P.A. (1981) Hydromechanical behavior of a deformable rock fracture subject to normal stress. *J. of Geophy. Res.*, **86**(B10), 9287–98.

Tsang, Y.W. and Witherspoon, P.A. (1983) The dependence of fracture mechanical and fluid flow properties on fracture roughness and sample size. *J. Geophys. Res.*, **88**(B3), 2359–66.

Turcotte, D.L. (1986) Fractals and fragmentation. *J. Geophys. Res.*, **92**, 1921–6.

Uhl, U.W. and Sharma, G.K. (1978) Results of pumping tests in crystalline rock aquifers. *Groundwater*, **16**, 167–76.

Vairogs, J.C.L., Hearn, D.W., Dareing, D.W. and Rhoades, V.W. (1971) Effect of rock stress on gas production from low permeability reservoirs. *J. Pet. Tech.*, **23**, 1161–7.

Van Poolen, H.D. and Hoover, D.B. (1970) Waste disposal and earthquakes at the Rocky Mountain Arsenal, Derby, Colorado, *J. Pet. Tech.*, pp. 983–93.

Vasilev, M.V., Zoteev, V.G., Morozov, V.N. (1978) Predicting the block size of the rocks of ore deposits. *Sov. Min. Sci.*, **13**, 557–61.

Vicsek, T. and Kertesz, J. (1981) Monte Carlo renormalization group approach to percolation on a continuum: test of universality. *J. Phys. A: Math. Gen.*, **14**, 131–37.

Vieth, D.L., Blanchard, M.B., Younker, J.L. and Voegele, M.D. (1985) In-situ testing from the exploratory shaft at Yucca Mountain, Nevada Test Site, Nevada. *Symposium on Coupled Processes, L.B.L.*

Wallis, P.F. and King, M.S. (1980) Discontinuity spacings in a crystalline rock. *Int. J. Rock Mech. Min. Sci.*, **17**, 63–6.

Walsh, J.B. (1981) Effects of pore pressure and confining pressure on fracture permeability. *Int. J. Rock. Mech. Min. Sci. & Geomech. Abstr.*, **18**, 429–35.

Walsh, J.B. and Brace, W.F (1984) The effects of pressure on porosity and the transport properties of rock. *J. Geophys. Res.*, **80**, 9425–31.

Watanabe, K. (1986) Stochastic evaluation of the two-dimensional continuity of fractures in a rock mass. *Int. J. Rock. Mech. Min. Sci. & Geomech. Abstr.*, **23**, 431–7.

Watson, B.P. and Leath, P.L. (1974) Conductivity in the two-dimensional site percolation problem. *Phys. Rev.*, **B9**, 4893–6.

Webster, D.S., Proctor, J.F. and Marine, I.W. (1970) Two Well Tracer Tests in Fractured Crystalline Rock, US Geol. Survey Water Supply Paper 1544-I.

Westerman, A.R., Reeves, M.J. and Attewell, P.B. (1982) Rock surface roughness and discontinuity closure at depth. *Proc. Symp. Strata Mechanics, Newcastle upon Tyne*, Elsevier, Amsterdam, pp. 63–8.

Whittaker, B.N. and Singh, R.N. (1978) Design aspects of barrier pillars against water-logged workings in coal mining operations. *Conf. on Water in Mining and Underground Works, Canada, Spain,* pp. 675–92.

Wilke, S., Guyon, E. and Marsily, G.D. (1985) Water penetration through fractured rock: test of a tridimensional percolation description. *Math. Geology,* **17**, 17–27.

Wilson, C.R. and Witherspoon, P.A. (1970) An Investigation of Laminar Flow in Fractured Porous Rock, May, University of California, Department of Civil Engineering, Berkeley, CA. Quoted in Ziegler, T.W., 1976.

Wilson, C.R., Witherspoon, P.A., Long, J.C.S. *et al.* (1983) Large-scale hydraulic conductivity measurement in fractured granite. *Int. J. Rock Mech. Min. Sci. & Geomech. Abstr.,* **20**, 269–76.

Witherspoon P.A., Wang, J.S.Y., Iwai, K. and Gale, J.E. (1980) Validity of cubic law for fluid flow in a deformable rock fracture. *Water Resources Res.,* **16**(6), 1016–24.

Young, A.P. and Stinchcombe, R.B. (1975) A renormalization group theory for percolation problems. *J. Phys. C: Solid State Phys.,* **8**, L535–L540.

Ziegler, T.W. (1976) Determination of Rock Mass Permeability, Technical Report S-76-2, US Army Corps of Engineer, Waterway Experiment Station, Vicksburg, MS.

Zoback, M.D. and Byerlee, J.D. (1976) Effect of high-pressure deformation on the permeability of Ottawa Sand. *Bull. Am. Assoc. Pet. Geol.,* **60**, 1531.

Index